ELECTRONICS AND TELECOMMUNICATIONS RESEARCH

TELECOMMUNICATIONS

APPLICATIONS, MODERN TECHNOLOGIES AND ECONOMIC IMPACT

ELECTRONICS AND TELECOMMUNICATIONS RESEARCH

ELECTRONICS AND TELECOMMUNICATIONS RESEARCH

TELECOMMUNICATIONS

APPLICATIONS, MODERN TECHNOLOGIES AND ECONOMIC IMPACT

JOSEPH P. BARRINGER
EDITOR

NOVINKA
New York

For permission to use material from this book please contact us:
Telephone 631-231-7269; Fax 631-231-8175
Web Site: http://www.novapublishers.com

Library of Congress Cataloging-in-Publication Data

Telecommunications : applications, modern technologies and economic impact / editor, Joseph P. Barringer.
pages cm
Includes index.
ISBN: 978-1-63117-141-3 (soft cover)
1. Telecommunication. 2. Telecommunication--Software. 3. Application software. 4. Communications software. 5. Telecommunication--Technological innovations. 6. Telecommunication--Economic aspects. I. Barringer, Joseph P.
TK5102.5.T39 2014
621.382--dc23

2013051373

Published by Nova Science Publishers, Inc. † New York

CONTENTS

PREFACE

In this book, the authors present current research in the study of telecommunications. Topics discussed include electronic communications services in the world of apps; computer algebra and symbolic processing in modern telecommunication applications; and explicit form of three novel classes of multiplier-less linear phase selective 2D FIR filter function.

Chapter 1 – The market for telecommunications networks and services is constantly changing. So is the regulatory approach. Technological progress creates new challenges for the creation of a level playing field between traditional and new operators and service providers. One of these challenges is the classification of "apps" in the regulatory context.

The present paper investigates this growing mobile service segment which, from an economic perspective, exists largely independently of the network operators. The increasing prevalence of smartphones in particular is making these services more and more popular. Network operators are thus not only facing increased competition in the segment populated by their own services, but also have to carry a rising volume of traffic on their technical infrastructures, while the economic return generated by the traffic largely passes them by. In addition, the providers of such applications position themselves as rivals (in some sub-segments) by offering services that compete with those of the network operators. Given that they do not follow the same "rules", as laid down by legislation, the existence or otherwise of a level playing field as regards competition is to be investigated. In the authors' view, the regulatory approach to this issue adopted thus far needs to be changed in order to maintain equality of competition among the various offerings and players operating at the different stages of the vertical value chain.

Chapter 2 – The task of the theory domain is to develop methods to solve practical problems, organizes and makes intelligible field of knowledge on which it is based. In order to solve them, it is needed to be considered as part of a logical structure, and they need to understand the structures which solve it.

The main tools, used in telecommunications for calculating, designing and analyzing, are based on the numerical-only algorithms. So, many computer-based numerical algorithms are developed as consequence of development of hardware and computer techniques in numerical mathematic, adapted to computer calculations, and the main task of numerical processing is construction of algorithms, but, the drawback is generating of a tremendous amount of numerical data, and the user might easily lose insight into the phenomenon being investigated, and numerical computation manipulates with numerical values, so, very often it is important to take care about final results: accuracy, execution time of numerical algorithm, efficiency-based reasoning etc.

Computer algebra system belongs to the contemporary trends that use very sophisticated algorithms combine expertise in many areas, such as dynamic system design, control engineering, and signal processing, and the user makes a better and more advanced troubleshooting. It is based on well-defined semantics and algebraic algorithms, it is used to simplify rational functions, find the solutions to a system of equations, and give general formulas as answers, model industrial mathematical problems, automated-proven properties and various other manipulations.

Symbolic processing is a special feature of these systems. That is the excitation signals may represent not only numbers, even the symbols, and it is analyzed what happens to them during processing. The parameters of the system can be analyzed subsequently appointing of additional requirements on the desired response, so, on this way the analysis is used in system design. Symbolic processing is concerned with alternative forms, partially-specified domains, and symbolic derivations in general and it can be treated as a transformation of expression trees, such as symbols for operations, variables, constants, and simplification, and plenty expressions, functions and equations can be derived in closed form using symbolic processing, as well as the error-free derived simulation code can be obtained.

Chapter 3 – The application of the extremal property of Christoffel-Darboux formula for classical continual orthonormal polynomials to the theory of synthesis of continual filter functions and digital multiplierless linear phase one-dimensional (1D) and two-dimensional (2D) FIR filter functions is completely studied and described by analytical methods, and the results are given in compact explicit form in literature.

By correct mapping from the continual domain to the z domain without errors and without residues, an ultra selective spike-like pass-band is obtained and unwanted parasitic Gibss effect is completely suppressed. In this chapter, we propose in explicit form three novel classes of multiplierless linear phase selective 2D FIR filter function, of symmetric order *N***N*, obtained directly by using analytical method in z domain. This paper represents introduction in ATLAS 2D FIR filter functions. Three novel classes of 2D FIR filter functions $H_i(N, z_1, z_2)$, $i \in (1,2,3)$ are generated by using cascade realization and by the derivation of two free real integer parameters the following form is obtained:

$$H_i(N, K, z_1, z_2) = [H_i(N-1, z_1, z_2) * H_i(N, z_1, z_2) * H_i(N+1, z_1, z_2)]^K, \quad i \in (1,2,3)$$

The examples of proposed filter functions are considered in the paper and the detailed analysis of amplitude response characteristics, magnitude response characteristics in dB and phase response characteristics is given. The shape and the width of pass-band of proposed novel classes of FIR filter function are studied in details for a given value of maximal attenuation in the filter pass-band. Also, the examples of stop-band of proposed ultra selective 2D FIR filter functions are illustrated. Parasitic Gibss effect is the main shortage of sloped interrupt of infinite Fourier order and for given examples it is evidently suppressed. The possible applications of these filters are for special telecommunication links as they have a very narrow pass-band and a good suppression in the stop-band. In addition, the proposed filters are of low complexity and with a high resolution and they are suitable for real-time applications. In addition, the proposed novel classes of 2D FIR filter functions can be very useful for future wireless applications and studies.

In: Telecommunications
Editor: Joseph P. Barringer

ISBN: 978-1-63117-141-3

Chapter 1

ELECTRONIC COMMUNICATIONS SERVICES IN THE WORLD OF APPS: REGULATORY CHALLENGES

Johannes Gungl[1], Ing. Mag. (FH), Alexander Gratzer[2], M.A, Dr. Ernst-Olav Ruhle[3] and Dr. Natascha Freund[3], LL.M

[1]Telecommunications Regulatory Authority (RTR), Austria
[2]Hutchison Drei Austria GmbH
[3]SBR Juconomy Consulting AG, Germany, Austria

ABSTRACT

The market for telecommunications networks and services is constantly changing. So is the regulatory approach. Technological progress creates new challenges for the creation of a level playing field between traditional and new operators and service providers. One of these challenges is the classification of "apps" in the regulatory context.

The present paper investigates this growing mobile service segment which, from an economic perspective, exists largely independently of the network operators. The increasing prevalence of smartphones in particular is making these services more and more popular. Network operators are thus not only facing increased competition in the segment populated by their own services, but also have to carry a rising volume of traffic on their technical infrastructures, while the economic return generated by the traffic largely passes them by. In addition, the providers of such applications position themselves as rivals (in some sub-segments) by offering services that compete with those of the network operators. Given that they do not follow the same "rules", as laid down by legislation, the existence or otherwise of a level playing field as regards competition is to be investigated. In the authors' view, the regulatory approach to this issue adopted thus far needs to be changed in order to maintain equality of competition among the various offerings and players operating at the different stages of the vertical value chain.

1. Introduction

Driven by technical and economic progress, the transformation of the telecoms markets is also exerting an influence on the business model adopted by classic network operators. For many years, networks were installed and operated, and services were offered to customers, by one and the same entity. Novel services were, as a rule, offerings of the network operators or tied to their technical infrastructure. But the market is no longer moulded by such a closed loop, or vertically integrated value chain. Nowadays, access to the market can be gained at any stage of the value creating chain.

From the customer's perspective as well, the world of communications has radically changed. Thanks to smartphones, customers have constant access to a computer. Irrespective of time and place, they can retrieve information from the internet, read messages and get in touch with friends and business partners. In contrast to the situation that prevailed just a few years ago, communication now takes place through a multitude of technologies – by telephone, text message (SMS), iMessage or MMS, through the messaging functions of social networks such as Facebook, and by way of messaging apps, including WhatsApp, Joyn, Viber, KakaoTalk, Skype, SkypeOut, Nimbuzz, NimbuzzOut and others.

On the touch screen of a smartphone, the buttons that give access to the telephone, messages, Nimbuzz, WhatsApp etc., differ only in appearance. In most cases, they stand alongside each other with equal status. Sometimes, the services of the terminal device's manufacturer are even integrated in the network operator's texting service so that the user does not immediately recognise which service he is using. In other words, before engaging in any communications activity, the customer can decide, or the terminal device decides on his behalf, which communications service to select.

The network operator's services are therefore competing directly with all the other communications services that are downloaded from the internet by the customer or pre-installed by the terminal device manufacturer. Only the access is still largely made available exclusively by the network operator. While services and applications on the internet are developed and distributed "globally", and, more to the point, device manufacturers serve the "world market" rather than national markets, another trend is emerging: this globalisation of the service and application markets is taking place within a structure for physical access, whether mobile or landline, that remains firmly national. But how are these applications to be regarded against the backdrop of national operators, markets and, in particular, national legislation? Are they subject to a national regulatory framework? To what extent does the Telecommunications Act (TKG) apply, and what are the implications of the new form of competition?

It is beyond doubt that these new applications are highly significant for the market. The comparative figures set forth below illustrate the forces that are at play. For June 2012, Facebook reported a total of 552 million daily active users,[1] which represented a year-on-year increase of 32%. Vodafone, one of the world's largest mobile providers, puts its global customer base at 404 million users.[2]

Viber, an instant messaging application for text and photo messages as well as for making HD-quality phone calls, reported 100 million customers in September 2012.[3] On 28 October 2012, the Nimbuzz Messenger, a messaging app for voice and video calls, chatting and transferring files, reported in its description in the App Store that it had 105 million registered users. KakaoTalk, a similar app, claims in the App Store that its customer portfolio stands at 62 million users in 230 countries (likewise as of 28 October 2012). According to a press release, WhatsApp customers send one billion messages a day.[4]

This competition is also reflected in the mobile companies' turnover figures. In some countries, their text messaging revenues are declining sharply. In the period from 2009 to 2011, text revenues fell by 10% in Spain and 2% in

[1] Facebook news release on second quarter 2012 results, http://www.prnewswire.com/news-releases/facebook-reports-second-quarter-2012-results-163915536.html.

[2] Vodafone Group Plc, Annual Report for the year ended 31 March 2012, page 3; http://www.vodafone.com/content/dam/vodafone/investors/annual_reports/Vodafone_Annual_Report_12.pdf.

[3] http://www.businesswire.com/news/home/20120911005071/en/Viber-Reaches-100-Million-Users-Takes-Aim (11.9.2012).

[4] http://www.whatsapp.com/press (28.10.2012).

Italy.[5] From October 2011 to August 2012, the use of text messaging among Finnish iPhone users dropped by 14%, while the use of mobile data in the same period rose by 68%.[6] In the period between Q2/2012 and Q2/2013 , SMS activity in Austria contracted from 2 billion to 1.54 billion messages.[7]

The forenamed applications are therefore clearly competing against and substituting the services provided by traditional mobile operators. Although no official figures are currently available for Austria or Germany, it is to be assumed – based on a smartphone penetration of 36% in Q1/2012, an average of 25 installed apps[8] and 2.14 million smartphone tariffs[9] – that many more than one million smartphone users in Austria were regularly sending messages by way of apps during this quarter. This situation affects network operators, who must be careful that their access offerings do not transform them into "bitpipes" – providers that "only" give access to networks, and leave the business of offering services to the application providers.

The present paper[10] investigates this growing mobile service segment which, from an economic perspective, exists largely independently of the network operators. The increasing prevalence of smartphones in particular is making these services more and more popular. The network operators are thus not only facing increased competition in the segment populated by their own services, but also have to carry a rising volume of traffic on their technical infrastructures,[11] while the economic return generated by the traffic largely passes them by. In addition, the providers of such applications position

---

[5] The Economist, Joyn them or join them, 11.8.2012, citing research firm Analysys Mason.

[6] http://www.forbes.com/sites/terokuittinen/2012/10/09/text-messaging-volume-of-iphone-owners-declining-sharply-in-finland/ (30.10.2012).

[7] RTR Telekom Monitor Q4/2013.

[8] Google, Our Mobile Planet May 2012, retrievable at www.http://www.thinkwithgoogle.com/mobileplanet (29.10.2012).

[9] RTR Telekom Monitor Q3/2012.

[10] The authors thank Mag. Jörg Kittl for his valuable comments on an earlier version of this paper.

[11] The signalling network, which enables consumers to use the mobile network by facilitating communication between it and its constituent elements on the one hand, and terminal devices on the other, is therefore being exposed to greater strain as well. It was found earlier, for example, that smartphones were generating 40% of the data traffic, but 99% of the signalling traffic, in a West European mobile network. This phenomenon is attributable to a feature called "fast dormancy", which is intended to prolong battery life. The terminal device interrupts the connection to the mobile network and enters the standby mode after each individual download of a data fragment. When an app is being used, this prompts the terminal device to establish and interrupt the connection to the mobile network extremely frequently (several times a second), however, and therefore imposes a burden on the network. Cf. Nokia Siemens Networks, Understanding smartphone behavior in the network, White Paper, p. 5.

themselves as rivals (in some sub-segments) by offering services that compete with those of the network operators. Given that they do not follow the same "rules", as laid down by legislation, the existence or otherwise of a level playing field as regards competition is to be investigated. At the same time, the Paper discusses the economic significance of this development and seeks to define its place within the legal and regulatory framework. In the authors' view, the regulatory approach to this issue adopted thus far needs to be changed in order to maintain equality of competition among the various offerings and players operating at the different stages of the vertical value chain.

After this Introduction, Chapter 2 describes some of the innovative services and categorises them within the portfolio of network operators' and application providers' offerings. Parallels and differences between the individual services are thus revealed. Chapter 3 classifies these offerings from a legal and regulatory perspective, and seeks to measure the applications against the yardstick of TKG 2003. The authors believe that the apps under consideration here are to be classified as communications services. The implications are described in Chapter 4, which precedes our conclusion in Chapter 5.

2. Parallels and Differences Between Network Operators' Services and Those of App Providers

2.1. Messaging Portals Becoming Mobile

As mobile phones become ever more popular, messaging portals and (instant) messaging systems are being accessed by mobile users to an increasing extent. And as the prevalence of smartphones grows, more and more apps are seeking to capture the space previously occupied by classic text messaging. In this climate, the boundaries between "social networks" and "communications services" are becoming more and more blurred.[12] In principle, all the systems operate in a very similar way.

[12] The "Facebook" social network, for example, in August 2011 launched the "Facebook Messenger", an app for the mobile exchange of messages between Facebook users; cf. Protalinski, E.: Facebook is killing text messaging, http://www.zdnet.com/blog/facebook/facebook-is-killing-text-messaging/12911.

Figure 1. Message history in the Facebook Messenger app.

By way of an application, messages can be sent to one or several platform users. Most platforms allow both text-only and text-with-picture messages. Figure 1 shows a typical message session, in this case taking place via the Facebook Messenger.

In view of these systems' large user numbers,[13] they are now taking the place of classic text message services.[14]

Two of the apps, WhatsApp and iMessage, are described below as examples of this genre.

---

[13] According to the German website "All Facebook", on 5 September 2012 there were 914,258,940 active Facebook users worldwide; in Austria alone, 2,794,300 users logged on with Facebook in the last 30 days (cf. http://allfacebook.de/userdata). However, much smaller platforms are generating appreciable demand as well; the kik messaging platform launched in April 2010, for instance, has already reported more than 19 million users (cf. http://kik.com/about, 5 September 2012).

[14] Pingdom put the number of instant messaging users in 2011 at around 2.6 billion; cf. http://royal.pingdom.com/2012/01/17/internet-2011-in-numbers, 5 September 2012. The messaging platform WhatsApp announced on 31 October 2011 that it had passed the global delivery milestone of 1 billion messages a day for the first time; cf. http://blog.whatsapp.com/index.php/2011/10/one-billion-messages, 5 September 2012.

2.2. WhatsApp

The WhatsApp Messenger uses an IP-based proprietary data protocol. The front end is an app, which is available for the most popular smartphone operating systems (iOS, Android, BlackBerry, Nokia and Windows Phone). According to WhatsApp, it was created "to build a better SMS alternative".[15]

WhatsApp operates rather like an instant messenger, but it addresses the called party by way of the E.164 telephone number. The user logs on with his own telephone number, which is verified by way of an SMS. The messages are then composed in an app, following more or less the same rules as a conventional text message.

The app examines the terminal device's telephone book and accepts all the entries whose numbers are registered with WhatsApp. An internet connection is required for the sending of messages to and from the WhatsApp server. Mostly this connection is provided by a mobile broadband service to which the user subscribes through his network operator.

At first sight, the app is little different from the mobile manifestations of classic instant messaging systems. One user can send to another (or to a predefined group of users) text messages, photos, videos and his current location. In 2013, voice messages were added to this portfolio. Messages are exchanged over the public internet.

Apart from the network operator's charges for using the internet, the messages are sent for free. iPhone users are currently charged a one-time fee of EUR 0.89 for downloading the app.[16] For the users of other operating systems, the software is free for the first year and can be used thereafter for an annual subscription of USD 0.99.[17]

WhatsApp differs from other portals and social networks, however, insofar as it uses the message recipient's general address. With other popular systems, the recipient must initially register with the messaging platform and either notify potential senders of his assigned identity or send them a message. WhatsApp, in contrast, addresses the recipient with his normal telephone number, which is linked to the relevant mobile phone on the WhatsApp server. WhatsApp is therefore able simply to use the addresses in the normal telephone book without saving an additional user name; in this respect, it operates exactly like a conventional text messaging service.

---

[15] http://www.whatsapp.com/about, 18 October 2012.

[16] Cf. https://itunes.apple.com/at/app/whatsapp-messenger/id310633997?mt=8, 21 November 2012.

[17] Cf. http://www.whatsapp.com/faq/de/general/20965922, 18 October 2012.

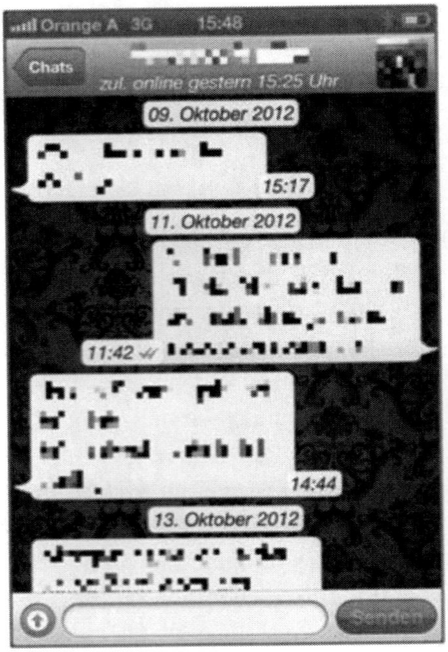

Figure 2. Message history in the WhatsApp Messenger app.

2.3. iMessage

Apple is pursuing a very similar concept with its iMessage service but made a significant advance when it launched the iOS 5.0 operating system and integrated the messenger in the mobile phone's existing SMS/MMS application.

Like WhatsApp, Apple links the relevant E.164 telephone number to the user's mobile phone number.[18] Here again, the telephone number is verified by way of a text message (a system SMS that is invisible to the user). In contrast to the WhatsApp approach, however, iMessage users do not need a separate application. Instead, the user composes his messages in the mobile phone's usual messaging app. If the device is connected to the internet when the user wishes to send the message, the app examines whether an iMessage connection exists for the recipient's phone number. If such a connection exists,

[18] Following the launch of OS X 10.8, iMessage can also be used with Macintosh computers. For this purpose, one or more e-mail addresses can be declared alongside the E.164 telephone number.

iMessage sends the message over the public internet. If there is no such connection, the message is sent as a conventional SMS or MMS by way of the network operator's SMSC or MMSC. Like WhatsApp, iMessage sends individual messages free of charge; the only costs are the network operator's fees for using the internet.

When sending a message, therefore, as a general rule the user does not consciously select a particular protocol (SMS/MMS, or iMessage over the public internet) for the communication.[19] From the user's perspective, iMessage is consequently no different from a conventional SMS or MMS service.

2.4. Commonalities and Differences in Apps

The forenamed applications thus primarily use the broadband access provided by the mobile operator. They are competing against the operators' own messaging and texting services.

From an economic perspective, therefore, the two applications described above are substituting the operators' SMS and MMS services. They differ inasmuch as one is marketed by a smartphone manufacturer, and the other by a third party, namely a service provider. Furthermore, the applications operate on foundations that differ from those of the "classic" services offered by network operators. These differences are outlined below.

- The service is generally used on an "as is" or "as available" basis. The conditions that have to be accepted by the user include liability exclusions, data protection provisions in some cases, as well as provisions concerning applicable law and jurisdiction (often outside the EU). As a result, the user often accepts conditions that are neither transparent nor compatible with the yardsticks that apply in the EU.
- Formerly, the purchase of a terminal device did not directly entail the use of services implemented by way of the device (unless in connection with a subsidised terminal device offering the network operator's own telephone services, typically for a fixed period). The situation is different nowadays. A smartphone user now utilises applications provided by the device manufacturer (or third parties) –

[19] The user does not have the option of making a direct choice – he can intervene only by completely deactivating the iMessage service in the mobile phone's settings.

in some cases such apps are pre-installed on the device – in combination with the network operator's mobile internet services.

- If the use is governed by standard business conditions, these are often formulated pursuant to the law of a third country, not consistent with Community law, and have not been examined for compliance with TKG 2003. Customer-specific data concerning purchasing behaviour in particular are collected and evaluated or, for example, processed to obtain movement profiles; the data can also be used by the provider for other commercial offerings. Given that users in many cases do not read or understand such standard business conditions, the public authority has a statutory duty to check them in advance.
- Against the background of this problem concerning consent to standard business conditions, the application of data and customer protection regulations is likewise not assured. As already indicated, such offerings are not subject to national scrutiny in many cases.
- Applications are developed for all users of certain smartphones. Since the market for devices is global, applications are also developed for the global market. A national outlook adopted from a legal/regulatory perspective, or by the network operators that implement the services, can give rise to contradictions and incompatibilities.

As a consequence, the offering of services and applications (by network operators and app providers) that are indistinguishable for the user as a general rule is taking place on economically diverse conditions. Application providers are subject to much "softer" regulation and, thanks to the more "relaxed" rules governing the treatment of users and their data, can garner competitive advantages and thus (further) strengthen their market position. The key questions in this context concerning the services and applications outlined above are: Can they be classified as electronic communications networks or services? What are the implications for a legal assessment of these services?

3. What Statutory Regulations Apply to Apps and "New Services"?

As regards the legal classification of the applications for statutory regulatory purposes, the Telecommunications Act and the E-Commerce Act could apply to the problem at issue. It is to be examined below whether the forenamed applications fall within the scope of the E-Commerce Act (ECG) and/or the TKG 2003. In particular, the issue raises questions concerning the extent to which the country-of-origin principle applies, and how this principle is to be interpreted in connection with TKG 2003. The specific question raised is: Does a service provider have to disclose its service in every country in which it operates, or is it sufficient for the service to be disclosed in the country in which the provider's head office is situated?

3.1. Directive 2000/31/EC in Conjunction with the ECG

The "Directive on electronic commerce"[20] was implemented in Austria by the enactment of the E-Commerce Act.[21]

The ECG provides a legal framework for certain aspects of electronic commerce and legal relations. It regulates the authorisation of service providers, their duties to provide information, the conclusion of contracts, the responsibility of service providers, the country-of-origin principle, and the cooperation with other member states in electronic commercial and legal transactions.[22]

Pursuant to Art. 3 No. 1 ECG, a messaging service is an information society service because, among other things, it transmits information via an electronic network.[23] The service provider and its service thus fall within the scope of ECG. The provider must therefore meet the obligations imposed by the Act and can also exercise the rights arising therefrom, including in particular the country-of-origin principle pursuant to Art. 20 ECG. As a consequence of this principle, the legal requirements to be satisfied by a service provider established in a member state are governed by the law of the state concerned. In other words, if the service provider complies with all the

[20] DIRECTIVE 2000/31/EC OF THE EUROPEAN PARLIAMENT AND OF THE COUNCIL of 8 June 2000 on certain legal aspects of information society services, in particular electronic commerce, in the Internal Market ("Directive on electronic commerce", Official Journal of the European Communities L 178/1 of 17.7.2000).

[21] Federal Law Gazette I No. 152/2001.

[22] Cf. Art. 1 (1) ECG.

[23] Art. 3 No. 1 ECG defines "information society service" as follows: a service normally provided in return for consideration electronically by distance selling at the individual retrieval request of the recipient (Art. 1 (1) No. 2 of the Notification Act of 1999), particularly the online marketing of goods and services, online information offers, online advertising, electronic search engines and data retrieval tools, as well as services that transmit information via an electronic network and provide access to such a network or store the information of a user.

statutory provisions of its country of origin, such as data and consumer protection regulations or provisions concerning the legal responsibility of service providers (Art. 3 No. 8 ECG), it cannot be made subject to stricter local rules. The services are to be supervised for all Community citizens at the source of the activity.[24]

An exception exists, however, that is highly relevant to our investigation. Pursuant to Art. 4 (2) ECG, legal provisions that regulate the authorisation for taking up or pursuing any trade, business or profession, but do not apply specifically and exclusively to information society services or the providers thereof, shall not be prejudiced. The same applies to legal provisions concerning the disclosure (notification) and concession duties for telecommunications services. The explanatory remarks[25] concerning Art. 4 (2) ECG accordingly make it clear that the E-Commerce Act, including the underlying principle excluding prior authorisation, shall not prejudice the notification and concession duties for certain telecommunications services (cf. Art. 13 and Art. 14 TKG 2003). This also corresponds to Directive 2000/31/EC insofar as its Art. 4 (2) rules that the principle excluding prior authorisation shall not prejudice authorisation schemes which are not specifically and exclusively targeted at information society services, or which are covered by Directive 97/13/EC of the European Parliament and of the Council of 10 April 1997 on a common framework for general authorisations and individual licences in the field of telecommunications services (Authorisation Directive).[26]

The first finding is, therefore, that pursuant to ECG, the country-of-origin applies as a general rule. Consequently, the law of the country in which the country has its head office is to be applied. Against the background of Art. 4 (2) ECG, if the applications are to be classified as telecommunications services, Austrian TKG is to be applied irrespective of the location of the relevant company's head office.

If this assumption is rejected, it could be argued that, although Art. 4 (2) ECG envisages the application in any event of TKG 2003 for telecommunications services as regards the classification of the apps that are at issue here, this applies exclusively to cases in which Austrian law applies because of the country-of-origin principle. It could thus be argued that a

[24] Cf. recital 22 of Directive 2000/31/EC, Official Journal of the European Communities L 178/4 of 17.7.2000.

[25] Explanatory remarks on government bill 817, annexes to the shorthand records of the National Assembly, 21st legislative period.

[26] Official Journal of the European Communities L 117/15 of 7.5.1997.

company offering telecommunications services over the internet is governed by the commercial and telecommunications law of the EU country of origin and therefore cannot be made subject to any additional requirements in other EU member states. This interpretation would allow such a service to be offered in Austria without disclosure. According to this construal, a point of contact with Austrian telecommunications law would depend on the EU provider developing activities in Austria that do not have to comply with TKG 2003. The available evidence gives too little indication of whether this particular outcome actually formed part of the legislative intent; after all, a service provider could opt to establish its head office and offer its services in and from a country whose regulations are "conducive" to its business. In turn, such a provider could entirely avoid the scope of application of TKG 2003 because only the regulations of its country of origin would apply, irrespective of the countries in which it offers its services and the regulations that apply there.

If the providers of these new services can be classified as operators of telecommunications services or networks, and the applications themselves are classified as telecommunications services, they are subject to all the obligations set forth in the Authorisation Directive within its scope of application. At this stage of the debate, the authors therefore believe that, in the case under consideration here, the Telecommunications Act, namely TKG 2003, applies as a general rule alongside ECG. Any other conclusion would both entail reverse discrimination (against domestic providers)[27] and contradict customary practice and the statutory requirement that providers of electronic communications services who are active in Austria must also give notice of their activity pursuant to Art. 15 TKG 2003. Against this background, it is to be examined which provisions of TKG 2003 apply in the case at issue here.

3.2. Austrian Telecommunications Act

Art. 3 No.3 TKG 2003 defines a communications service operator as an undertaking which exercises legal control over the functions in their entirety that are needed to provide the respective communications service and which offers the service to others. It is to be assumed that a provider of applications exercises legal control over the functions in their entirety that are needed to

[27] Otherwise, every provider would select a country having "favourable" commercial law provisions, establish itself there and then offer services in every EU member state without being subject to the regulations that apply in each individual country.

provide the respective communications service. This assumption is likely to hold even if the messages are routed via the internet because – as set forth in Chapter 2 – verification takes place by way of a text message in the case of these applications as well. If the existence of legal control were to be rejected, text messaging would presumably likewise not qualify as a communications service. The Austrian regulatory authority does, however, recognise text messaging as a communications service.

A messaging service can be compared with a VOIP product. In this context, the guidelines of the Austrian regulatory authority indicate that a service shall qualify as a communications service even if the operator only provides the called IP address.[28] The infrastructure extends even further, however, in the case of messaging services – messages are transmitted via the provider's server, where they are also apparently buffered. This supports the assertion that a service is being operated.

In the case of VOIP services, by analogy, all providers qualify as service providers if they give access to the public network by way of numbers belonging to the national or international numbering plan (publicly available telephone service, PATS). WhatsApp and Apple in particular address the called party by way of an E.164 telephone number. This also supports the argument that the provider is an operator. Although the app provider does not possess legal sovereignty over the entirety of the transmission, nor does a classic VOIP provider, who is nonetheless classified as a service provider. An equal evaluation of services and providers therefore makes it necessary to classify applications as the operating of a telecommunications service.

In summary, the conditions of Art. 3 No. 3 TKG 2003 are satisfied and the qualification as an operator can therefore be affirmed; the extent to which the conditions governing the existence of a communications service are satisfied is to be examined below.[29]

---

[28] Cf. RTR (Austrian Regulatory Authority for Broadcasting and Telecommunications): RichtlinienfürAnbieter von VoIP Diensten, p. 4 ff, see https://www.rtr.at/de/tk/RichtlinienVoIP.

[29] The definitions contained in the German Telecommunications Act (dTKG) are as follows: (1) Pursuant to Art. 3 No. 24 dTKG, "telecommunications services" means services normally provided for remuneration consisting in, or having as their principal feature, the conveyance of signals by means of telecommunications networks, and includes transmission services in networks used for broadcasting. This definition differs from the one enshrined in Austrian law only to the extent that the Austrian TKG contains the word "commercial" instead of "for remuneration", and that the defined term itself is "telecommunications services", rather than "communications services" as in Austrian TKG 2003; (2) Art. 3 No. 27 dTKG defines "telecommunications network" as transmission systems and, where applicable, switching and routing equipment and other resources, including the inactive network components, in

Whether, and the extent to which, the relevant applications constitute communications services in the meaning of TKG 2003 is defined in Art. 3 No.9 TKG 2003. "Communications service" is defined there as a commercial service which consists wholly or mainly in the conveyance of signals on communications networks, including telecommunications services and transmission services in networks used for broadcasting, but excluding services providing, or exercising editorial control over, content transmitted using communications networks and services. The term does not include information society services, as defined in Art. 1 (1) No.2 of the Notification Act [Notifikationsgesetz], Federal Law Gazette I No.183/1999, which do not consist wholly or mainly in the conveyance of signals on communications networks. On the one hand, the offering of apps is a commercial service,[30] and on the other hand, signals are conveyed by way of communications networks. If for no other reason, these claims are substantiated because the user, as in a text message, determines the content. It follows that the provider does not have a duty to perform any content monitoring whatsoever and, therefore, has no opportunity whatsoever to censor "impermissible" content. The defining characteristics of "communications service" pursuant to Art. 3 No.9 TKG 2003 are thus satisfied by the apps under consideration.

It is to be examined whether the provided apps are ancillary services. If they were deemed to fall within the scope of the term, the apps would consequently be excluded from the definition contained in Art. 3 No. 9 TKG 2003.[31] Although the underlying performance would as a general rule be

---

their entirety which permit the conveyance of signals by wire, by radio, by optical or by other electromagnetic means, including satellite networks, fixed and mobile terrestrial networks, electricity cable systems to the extent that they are used for the purpose of transmitting signals, networks used for radio and television broadcasting, and cable television networks, irrespective of the type of information conveyed. Here again, the definition is very similar to the one contained in Austrian TKG 2003, and differs only as regards the defined term ("communications network" in Austrian TKG, "telecommunications network" in dTKG). The public communications network mentioned in Art. 3 No. 17 TKG 2003 (a communications network used wholly or mainly for the provision of publicly available communications services) also exists in dTKG as "public telecommunications network", which is defined therein as a telecommunications network used wholly or mainly for the provision of publicly available telecommunications services that allow the transmission of information between network termination points. In view of the practically identical definitions, the classification of applications pursuant to the German Telecommunications Act must correspond to their classification pursuant to Austrian TKG 2003.

30 Remuneration is not an essential defining characteristic of the term "commercial", which means "of, connected with or engaged in economic activity".

31 Cf. explanatory remarks on government bill 128, annexes to the shorthand records of the National Assembly, 22nd legislative period. The argument concerning an ancillary service

regarded as the reselling of a communications service, it would constitute only a small part of another contract having a different content and being at variance with the main purpose of a communications service (e.g. accommodation, provision of office premises, industry-overarching customer loyalty programmes, regionalised marketing services), whereas such contract would not consist wholly or mainly in the conveyance of signals on communications networks. Not least against the background of the increasing use of these applications, there are no arguments to substantiate the claim that they are ancillary services, in particular, among other reasons, because they are not a small part of another contract having a different content and being at variance with the main purpose of providing a communications service, and whose content is wholly or mainly different from the conveyance of signals on communications networks.

This reflects the reasoning of the Telecoms Control Commission when it ruled that Skype Communications S.a.r.l. was obliged to give notice of its Skype Out service pursuant to Art. 15 TKG 2003. In this light, there are no evident arguments against the classification of the apps described here as communications services, as defined by TKG 2003, which are to be disclosed to the regulatory authority pursuant to Art. 15 TKG 2003.

4. Consequences

The discussion outlined in Chapter 3 has revealed a lack of clarity in relation to both the application of the country-of-origin principle pursuant to ECG and the additional requirements of TKG 2003. If, in view of the globalisation of the app and terminal device markets, the country-of-origin principle were to be invoked, a provider would be able to circumvent the provisions of TKG 2003 even if it developed activities in Austria. Hereinafter, the authors describe the regulatory implications of this scenario; in other words, the consequences that would arise if the relevant applications were classified as communications services.

For undertakings that are to be classified as operators of communications networks or providers of communications services, the provisions of TKG 2003 set forth below come into question. The list that follows contains the

was put forward by Skype Communications S.a.r.l. in relation to its Skype Out service in the procedure R 8/08 before the Telecoms Control Commission. The regulatory authority found, however, that the assertion lacked substantiation. Cf. in this connection the particulars of the Telecoms Control Commission's decision R 8/08 of 23.6.2008.

provisions that can be regarded in the broadest sense as "duties" of network operators and service providers. According to the findings of Chapter 3, application providers are therefore subject to these duties and requirements as well:

- Notification obligation (Art. 15 TKG 2003)
- Security and integrity (Art. 16a TKG 2003)
- Service quality (Art. 17 TKG 2003)
- Terms and conditions as well as tariffs (Art. 25 TKG 2003)
- Special information obligations (Art. 25 b TKG 2003)
- Financial contribution (Art. 31 TKG 2003 and Art. 34 KOG (KommAustria Act))
- Market analysis/SMP/remedies (regulation of competition pursuant to Art. 35 et seq.TKG 2003)
- Duties to provide information (Art. 90 TKG 2003)
- Data protection (Art. 92 to Art. 97 TKG 2003)[32]

"Classic" network operators are subject to these provisions of TKG. Material provisions among these are briefly illuminated below:

a) Notification obligation (Art. 15 TKG 2003)
First, pursuant to Art. 15 TKG 2003,

> "The intended provision of a public communications network or service as well as its modifications and its termination shall be notified to the regulatory authority prior to the start of operation, modification or termination."[33]

To date, companies that offer such apps or similar services, including Google,[34] Apple, Amazon, WhatsApp and others, have not submitted any notification pursuant to Art. 15 TKG 2003.

---

[32] Data retention pursuant to Art. 102a TKG 2003 would also come into question as a general rule, but the applications are outside the scope of services for which data retention is to be practised.

[33] The notification must indicate the name and address of the provider, the legal structure of the undertaking, where appropriate, a short description of the network or service, and the anticipated date of the start of operation, modification or termination of the service.

[34] This notification has to be submitted irrespective of the role played by the provider in the market. An entirely different subject under discussion is the possibly existing market dominance of Google in certain online and search engine segments, cf. for example

b) Security and integrity (Art. 16a TKG 2003)
Art. 16a TKG 2003 requires operators of public communications networks and services to guarantee the security and integrity of their networks. Among the steps cited in this connection are:

"reasonable technical and organisational measures"

as required to control risks to network security. Information concerning such measures must be provided to the regulatory authority. Security breaches and integrity losses are to be reported to the regulatory authority.[35] In such cases, obligations can be imposed, including the requirement to submit to a security audit at the operator's expense (Art. 16a (4) TKG 2003). If a service is not disclosed, the authority is unlikely to be able to make such demands and insist that information is furnished. Operators of public communications networks and services that do not meet their notification obligation are therefore not monitored by the authority.

c) Service quality (Art. 17 TKG 2003)
Pursuant to Art. 17 (1) TKG 2003, operators of public communications services shall

"publish comparable, adequate and up-to-date information on the quality of their services and on the measures taken to ensure equivalence in access to publicly available telecommunications services for users with disabilities, and provide the regulatory authority with this information at its request prior to publishing."

Applications constituting economic substitutes for existing services would have to be subject to this provision as well. No such published requirement is to be found, however, and, if a service is not disclosed, the regulatory authority does not submit any such request to the relevant provider. It is worthy of note that the quality assurance requirements affect operators of public communications services, even if they implement their service by way of a third-party network.

Rosenbach, M.: "Google istextrem dominant" and Müller, M.U. et al: "Der Türsteher", both articles in Der Spiegel No. 43/2012, p. 90 and p. 86 ff respectively.

35 The numerous security problems associated with the WhatsApp service, for example, have been well publicised (typical of many articles is http://www.20min.ch/digital/news/story/-Die-Probleme-bei-WhatsApp-sind-Tatsache--24514892), but, as far as is known, the regulatory authority has not taken any counteraction.

d) Terms and conditions as well as tariffs (Art. 25 TKG 2003)
Art. 25 TKG 2003 contains provisions governing standard business conditions (terms and conditions; AGB) and tariff conditions (EB).Art. 25 (1) TKG 2003 states that operators of communications networks or services

"shall issue general terms and conditions which shall also comprise a description of the services, as well as define the relevant tariff conditions. General terms and conditions as well as tariffs shall be notified to the regulatory authority before provision of the service is started and shall be promulgated in an appropriate form."

In addition, Art. 25 (2) TKG 2003 stipulates that

"Changes in general terms and conditions as well as in tariffs shall be notified to the regulatory authority before they take effect and shall be promulgated in an appropriate form."

Art. 25 (4) TKG 2003 governs the requirements that apply to terms and conditions. Here again, the terms and conditions of application providers are not subject to such scrutiny by the regulatory authority. Their compliance with Austrian law is therefore not ensured.

e) Universal service (Art. 26 et seq.TKG 2003)
TKG 2003 contains extensive provisions concerning universal service. According to these provisions,

"The provider of universal service shall be compensated, at his request, for the established costs incurred by the provision of the service that are unrecoverable despite efficient management if these costs constitute an unreasonable burden."(Art. 31 (1) TKG 2003)

Pursuant to Art. 32 TKG 2003, if required, the regulatory authority shall set up and administer a Universal Service Fund for the financing of universal service (Art. 31 (1) TKG 2003). Para.2 rules that operators of telecommunications services with an annual turnover from these activities of more than € 5 million shall contribute to the financing of the Universal Service Fund and to the financing of the fund administration in proportion to their market shares (universal service charge). The share shall be determined according to the proportion of their turnover to the total turnover of the companies assessed as liable to contribute on the relevant product market in Austria. In the absence of relevant data, whether and the extent to which application providers generate turnover of more than € 5 million cannot be assessed. If such providers were generating revenues in this amount, they would, by omitting to disclose their services, be evading their financing obligation and thus obtaining a benefit at the expense of providers that disclose their services. Given that for many years in Austria a universal service deficit of A1 Telekom Austria was mitigated by other providers' payments, the apportionment of these payments to the individual providers on the Austrian market is absolutely relevant.

f) Duties to provide information (Art. 90 TKG 2003)

Art. 90 TKG 2003 governs the reporting requirements imposed on the operators of communications networks and services, according to which the regulatory authority is to be furnished with extensive information. Here again, in the current situation, application providers that do not disclose their networks and services de facto avoid these duties, while "classic" network operators and service providers have to submit the following and other items of information:

- information required for the systematic or individual review of the obligations arising from this Federal Act or an ordinance or official decision issued under this Federal Act;
- information required for the individual review of the obligations if a complaint has been addressed to the regulatory authority, or if it assumes a violation of duties for other reasons or conducts investigations on its own behalf;
- information in procedures for the licensing of frequencies or communications parameters;
- information required for a procedure pursuant to Arts.36 to 37a;
- information required for the publication of quality and price comparisons for services to the benefit of consumers; and
- information on future network or service developments which may affect the relevant services existing at the wholesale level.

g) Supervision measures (Art. 91 TKG 2003)

The regulatory authority also disposes of supervision measures pursuant to Art. 91 TKG 2003, which it can impose on operators that must comply with TKG 2003. Here again, an unregistered provider can therefore evade the measures which the regulatory authority can impose pursuant to Art. 91 TKG 2003.

h) Data protection (Art. 92 et seq.TKG 2003)
Data protection and data security are significant issues for many applications. This claim is underpinned by questions concerning the security of applications such as WhatsApp (see above). In this connection, Art. 95 TKG 2003 imposes special data security requirements on every operator of a public communications service for each service provided by that operator. Art. 95 (2) TKG 2003 rules that

> "[...] in case of a particular risk of a violation of confidentiality, the operator of a public communications service must inform the subscribers about such risk and, where the risk lies outside the scope of the measures to be taken by the operator, about any possible remedies including their costs."

This requirement is not being met by application providers.
Later, Art. 95a TKG 2003 contains provisions on security breaches. Pursuant to Para.2, in the event of a breach of the security of personal data,

> "[...] the provider of public communications services shall, without delay, notify the Austrian Data Protection Commission of the personal data breach. In cases where such a breach is likely to adversely affect the privacy or personal data of individuals, the operator shall also notify the individuals affected by the breach without delay."

Para.6 requires that operators of public communications networks maintain an inventory of personal data breaches. This inventory must comprise the facts surrounding the breaches, their effects and the remedial action taken, and must be sufficient to enable the Data Protection Commission to verify compliance with the provisions of Paras.1 to 4. As in other instances, undertakings that have disclosed their services comply with these obligations; application providers, in contrast, avoid the work and expense.

i) Financing of RTR-GmbH (Art. 34 KOG)
The telecommunications industry has to pay financial contributions in order to finance the expenses of RTR-GmbH (Austrian Regulatory

Authority for Broadcasting and Telecommunications) incurred in the performance of its duties concerning the industry. The telecommunications industry encompasses the providers that are subject to the notification obligation of Art. 15 TKG 2003.[36] The financial contributions are assessed and charged according to the proportion of the liable undertaking's turnover to the industry's total turnover (Art. 14 (3) KOG). The budgeted expenditure of RTR-GmbH for the telecommunications industry in 2012 is € 7.501 million.[37]

The forenamed provisions give rise to substantial work and expenditure on the part of the relevant undertakings in some cases, for human resources, technical implementation, and the implementation of processes and systems. If application providers perform services that are notifiable pursuant to TKG 2003, but omit to comply with their notification obligation, they thus obtain the following competitive advantages:

- The quality of their services is not monitored. The regulatory authority is unable to monitor compliance with quality yardsticks or take exception to defects; quality complaints raised by consumers "fall on deaf ears".
- The terms and conditions of these companies are often accepted "implicitly" when customers use the service. The compliance of the terms and conditions with TKG 2003 is not examined.
- In the absence of an opportunity for the relevant provisions of TKG 2003 to be applied, these undertakings are not obliged to furnish the regulatory authority with information. Their costs are therefore lower than those of providers that collect and furnish the relevant information in compliance with the TKG; and this situation reduces the market's transparency.
- The companies do not make any financial contribution to the Universal Service Fund Network operators and service operators that have disclosed their operations therefore pay an elevated contribution and incur a competitive disadvantage.

[36] Art. 34 KommAustria Act (KOG), Federal Law Gazette I No. 32/2001, as amended by Federal Law Gazette I No. 125/2011.

[37] Cf. public consultation document of RTR-GmbH on the 2013 budget for telecommunications, post and media regulation (only in German), published on 19.11.2012, p. 5, retrievable from www.rtr.at.

- The regulatory authority is financed only by companies that have disclosed their operations. Application providers avoid this expense.
- The undertakings do not take part in the market analysis proceedings, which can give rise to a distorted picture of market forces and offerings.
- Customers cannot invoke consumer protection standards. The same deficit applies as regards data protection.
- Compliance with provisions concerning data security and security breaches is not examined, and these provisions are possibly avoided.

These circumstances lead to the conclusion that the app providers, by omitting to disclose the operation of electronic communications networks and the offering of electronic communications services, obtain a considerable financial and non-financial advantage over undertakings that have complied with their notification obligation and thus satisfy the requirements of TKG 2003.[38]

The lack of regulation of the app providers therefore impairs competition and imposes disadvantages on both the companies competing with such

---

[38] In relation to Germany as well, the question arises as to which statutory provisions are applied to the providers of telecommunications services and networks. The following is a brief summary of the relevant provisions: (1) participation in the procedures of market regulation (market definition, market analysis and, if applicable, market dominance), Art. 9 et seq. dTKG; (2) interconnection undertaking (applies to operators of public telecommunications networks), Art. 16 et seq. dTKG; (3) if the existence of significant market power is established, ex ante obligations can be imposed, e.g. non-discrimination, transparency, access, price control; (4) compliance with the net neutrality requirement, Art. 41 a dTKG; (5) customer protection requirement, Arts. 43 a, 43 b, 44, 44 a and 45 a-p dTKG, depending on the stipulation and customer protection regulation, in some cases these provisions apply to providers of public telecommunications services, telecommunications services or publicly available telecommunications services; (6) provisions concerning numbering, including price indication, announcement and display, price caps etc. pursuant to Arts. 66 and 66 a-m dTKG; (7) if applicable, contribution to financing the universal service pursuant to Arts. 78-87 dTKG; (8) data protection provisions (Art 91 et seq. dTKG): Worthy of particular stress here are, for example, the duties to provide information pursuant to Art. 93 dTKG; (9) data security (Art. 109 a dTKG, applies to providers of publicly available telecommunications services); (10) implementation of intercepts, provision of information (Art. 110 dTKG). In this connection, the provisions of Arts. 111 and 112 dTKG also apply in relation to data for information seeking security authorities and the automated information procedure; (11) information requests (Art. 127 dTKG). Contraventions of, or non-compliance with, the requirements of the Telecommunications Act are punishable by fines pursuant to Art. 149 dTKG. Both the contraventions and the amount of the fines for administrative offences are defined in Art. 149 dTKG.

providers and the public sector, which is unable to guarantee compliance with its statutory duties.

CONCLUSION

The preceding analysis has shown that providers of applications are to be classified as operators of electronic communications services as per TKG 2003 at least insofar as they offer services that constitute economic substitutes for the network operators' own services. The privilege of the country-of-origin principle enshrined in the E-Commerce Act does not apply in this instance, so that the notification obligation pursuant to Art. 15 TKG 2003 becomes effective. If this were not the case, the outcome would hand a competitive advantage to foreign providers and discriminate against domestic providers by imposing only upon them the more rigorous obligations pursuant to TKG 2003.

The consequence of this argument is that app providers too are subject to the provisions of TKG 2003. To date, it has not become the established practice for these providers to disclose the services to the regulatory authority, so that their activities are out of line with those of undertakings that have properly disclosed the operation of electronic communications networks and services. By omitting to give notification, these app providers enjoy numerous competitive advantages, manifest either in lower costs, or in their own definition of "rules", in particular in their relationship with customers, that disregard consumer protection standards. Unequal regulatory treatment gives rise to massive economic benefits for the application providers, both directly in respect of economic (financial) advantages, and indirectly inasmuch as their less stringent regulation reduces the undertakings' direct and indirect costs.[39]

[39] The scope of this debate extends beyond apps, as illustrated by an article in the German daily Frankfurter AllgemeineZeitung of 8.2.2013, p. 7, in which the author questions whether access regulation, as envisaged by the (German) Telecommunications Act can be applied to social networks. He writes, "According to Article 16 of the Telecommunications Act (TKG), every telecommunications provider shall, upon request, undertake to make an interconnection offer to other providers in order to secure user communication, the provision of telecommunications services and service interoperability. Pursuant to Article 21 TKG, the regulatory authority can even directly compel providers having significant market power to interconnect their networks. The reason for these provisions – which originate in the requirements of European law and exist in a similar form in the United States as well – is the concern that telecommunications providers could be unwilling voluntarily to allow up-and-coming providers access to their networks. Articles 16 and 21 TKG create the conditions in which an O2 customer can contact a T-Mobile customer and

The refusal of an application provider to observe the provisions of TKG 2003 can give rise to the imposition of administrative penalties. Some of these penalties are as follows:

- Pursuant to Art. 109 (4) No.1 TKG 2003, a provider that fails to fulfil the notification obligation and is thus guilty of an administrative offence can be punished by a fine of up to € 58,000.
- An administrative fine in the same amount can be imposed if terms and conditions are not disclosed pursuant to Art. 25 TKG 2003.
- A fine of up to € 37,000 can be payable if the obligation to take steps in order to ensure security and integrity is infringed (Art. 16 a TKG 2003).
- In the event of non-compliance with the provisions concerning service quality, the regulatory authority can impose a fine of up to € 37,000.

In view of the applications' classification as electronic communications networks and services as per TKG 2003, competitive equality has to be established. For this purpose, providers of such networks and services must be subject to the provisions of TKG 2003. As a first step, the authority must therefore call on the app providers to comply with the notification obligation pursuant to Art. 15 TKG 2003 concerning the operation of electronic communications services and networks, If such action proves fruitless, appropriate administrative penalties are to be established and implemented.

Competitive equality implies that – insofar as the network operators are required to observe neutrality as regards the services they carry ("net neutrality") and are not allowed to treat the Facebook service preferentially to the detriment of WhatsApp, for example – at the same time, the network operators are to be treated with "regulatory neutrality", so that their services are afforded equal rights by the regulatory regime and can therefore be performed under equivalent economic conditions. The current situation gives rise to discrimination and distorts competition. It can be resolved by adopting

---

vice versa. As a consequence, customers sign up to the provider that offers them the best conditions, not necessarily the provider with the most subscribers. If the same principles were applied to virtual social networks, the basic functions of the existing (and new) platforms would have to be (made) interoperable. In social networking, the activity corresponding to the making of calls in the mobile phone market would be in particular – but not exclusively – making contact with individuals or several persons by text. In order to make such contact possible from any platform to the users of other platforms, a unique address – accessible from every platform – could be assigned to the profiles in social networks."

one of two approaches. The first entails granting "regulatory freedom" to classic services in order to facilitate the free play of forces on the market populated by network operators and application providers. In view of the history of telecommunications regulation in Europe, this solution appears to offer little likelihood of success. The other solution requires the regulation of the application providers according to the provisions of telecommunications law; in other words, the establishment of a "level playing field" based on telecommunications law.

ACKNOWLEDGMENTS

We thank Mag. Kittl, also of SBR Juconomy Consulting AG, for his valuable comments on the present White Paper. Mag. Johannes Gungl will resume the position as head of the Austrian regulatory authority for telecommunications (RTR GmbH) as of February 2014.

REFERENCES

Directive 2000/31/EC of the European Parliament and of the Council of 8 June 2000 on certain legal aspects of information society services, in particular electronic commerce, in the Internal Market ("Directive on electronic commerce", *Official Journal of the European Communities* L 178/1 of 17.7.2000).

dTKG: German Telecommunications Act (*Telekommunikationsgesetz,* TKG).

ECG: Austrian E-Commerce Act (E-Commerce-Gesetz).

Federal Law Gazette I No. 32/2001.

Federal Law Gazette I No. 125/2011.

Federal Law Gazette I No. 152/2001.

http://allfacebook.de/userdata (Retrieved September 5, 2012).

http://blog.whatsapp.com/index.php/2011/10/one-billion-messages (Retrieved September 5, 2012).

http://kik.com/about (Retrieved September 5, 2012).

http://royal.pingdom.com/2012/01/17/internet-2011-in-numbers (Retrieved September 5, 2012).

http://www.20min.ch/digital/news/story/-Die-Probleme-bei-WhatsApp-sind-Tatsache--24514892 (Retrieved September 5, 2012).

http://www.businesswire.com/news/home/20120911005071/en/Viber-Reaches-100-Million-Users-Takes-Aim (Retrieved September 11, 2012).

http://www.forbes.com/sites/terokuittinen/2012/10/09/text-messaging-volume-of-iphone-owners-declining-sharply-in-finland/ (October 30, 2012).

http://www.prnewswire.com/news-releases/facebook-reports-second-quarter-2012-results-163915536.html (Retrieved September 2012).

http://www.thinkwithgoogle.com/mobileplanet (Retrieved October 29, 2012).

http://www.whatsapp.com/about (Retrieved October 18, 2012).

http://www.whatsapp.com/faq/de/general/20965922 (Retrieved October 18, 2012).

http://www.whatsapp.com/press (Retrieved 28 October 2012).

https://itunes.apple.com/at/app/whatsapp-messenger/id310633997?mt=8 (Retrieved November 21, 2012).

KommAustria Act (KOG). Retrieved October 2012 from https://www.rtr.at/en/m/KOG.

Nokia Siemens Networks Smart Labs: Understanding smartphone behavior in the network, *White Paper*, 2011, p. 5.

Official Journal of the European Communities L 117/15 of 7.5.1997.

Protalinski, E.: Facebook is killing text messaging. Retrieved October 2012 from http://www.zdnet.com/blog/facebook/facebook-is-killing-text-messaging/12911.

Rosenbach, M.: "Google istextrem dominant" and Müller, M.U. et al: "Der Türsteher", both articles in: *Der Spiegel* No. 43/2012, p. 90 and p. 86 ff respectively.

RTR (Austrian Regulatory Authority for Broadcasting and Telecommunications): *RichtlinienfürAnbieter von VoIP Diensten*, p. 4 ff. Retrieved October 2012 from https://www.rtr.at/de/tk/RichtlinienVoIP.

RTR Telekom Monitor Q3/2012.

Telecoms Control Commission's decision R 8/08 of 23.6.2008.

The Economist: Joyn them or join them (2012). Retrieved September 2012 from http://www.economist.com/node/21560298.

TKG 2003.

Vodafone Group Plc: *Annual Report for the year ended 31 March 2012*, page 3. Retrieved October 2012 fromhttp://www.vodafone.com/content/dam/vodafone/investors/annual_reports/Vodafone_Annual_Report_12.pdf.

In: Telecommunications
Editor: Joseph P. Barringer

ISBN: 978-1-63117-141-3

Chapter 2

Computer Algebra and Symbolic Processing in Modern Telecommunication Applications: A New Kind of Survey

Dr Vlastimir D. Pavlovic[1*], Full professor,
Dr Vladimir Mladenovic[2†], Assistant professor
and Dr Miroslav Lutovac[3‡], Full professor
[1]Department of Electronics, Faculty of Electrical Engineering,
University of Nish, Nish, Republic of Serbia
[2]Department of Information Technologies, Faculty of Technical Sciences
in Cacak, University of Kragujevac, Cacak, Republic of Serbia
[3]Faculty of Technical Sciences, Singidunum University,
Belgrade, Republic of Serbia

Abstract

The task of the theory domain is to develop methods to solve practical problems, organizes and makes intelligible field of knowledge on which it is based. In order to solve them, it is needed to be considered

* email: vlastimir.pavlovic@elfak.ni.ac.rs.
† email: vladimir.mladenovic@ftn.kg.ac.rs.
‡ email: lutovac@gmail.com.

as part of a logical structure, and they need to understand the structures which solve it.

The main tools, used in telecommunications for calculating, designing and analyzing, are based on the numerical-only algorithms. So, many computer-based numerical algorithms are developed as consequence of development of hardware and computer techniques in numerical mathematic, adapted to computer calculations, and the main task of numerical processing is construction of algorithms, but, the drawback is generating of a tremendous amount of numerical data, and the user might easily lose insight into the phenomenon being investigated, and numerical computation manipulates with numerical values, so, very often it is important to take care about final results: accuracy, execution time of numerical algorithm, efficiency-based reasoning etc.

Computer algebra system belongs to the contemporary trends that use very sophisticated algorithms combine expertise in many areas, such as dynamic system design, control engineering, and signal processing, and the user makes a better and more advanced troubleshooting. It is based on well-defined semantics and algebraic algorithms, it is used to simplify rational functions, find the solutions to a system of equations, and give general formulas as answers, model industrial mathematical problems, automated-proven properties and various other manipulations.

Symbolic processing is a special feature of these systems. That is the excitation signals may represent not only numbers, even the symbols, and it is analyzed what happens to them during processing. The parameters of the system can be analyzed subsequently appointing of additional requirements on the desired response, so, on this way the analysis is used in system design. Symbolic processing is concerned with alternative forms, partially-specified domains, and symbolic derivations in general and it can be treated as a transformation of expression trees, such as symbols for operations, variables, constants, and simplification, and plenty expressions, functions and equations can be derived in closed form using symbolic processing, as well as the error-free derived simulation code can be obtained.

I. Introduction

The task of the theory domain is to develop methods to solve practical problems, organizes and makes intelligible field of knowledge on which it is based. In order to solve them, it is needed to be considered as part of a logical structure, and they need to understand the structures which solve it.

The main tools, used in telecommunications for calculating, designing and analyzing, are based on the numerical-only algorithms. So, many computer-based numerical algorithms are developed as consequence of development of hardware and computer techniques in numerical mathematic, adapted to computer calculations, and the main task of numerical processing is construction of algorithms, but, the drawback is generating of a tremendous amount of numerical data, and the user might easily lose insight into the phenomenon being investigated, and numerical computation manipulates with numerical values, so, very often it is important to take care about final results: accuracy, execution time of numerical algorithm, efficiency-based reasoning etc.

A new concept provides advanced solutions to all the existing gaps and pushes the boundaries to new models of survey and design. The primary aspiration is speeding up in design and analysis of low-complexity implementation into high-complexity structures, which are provided by significantly increasing of processor's power and higher storage space, so the expert-based knowledge is necessary in areas such as theory of systems, signal processing, telecommunications, and software engineering.

Computer algebra system belongs to the contemporary trends that use very sophisticated algorithms combine expertise in many areas, such as dynamic system design, control engineering, and signal processing, and the user makes a better and more advanced troubleshooting. The computer algebra system finds a very efficient application, particularly in implementation of linear/non-linear systems, modeling, and simulation. It is based on well-defined semantics and algebraic algorithms, it is used to simplify rational functions, find the solutions to a system of equations, and give general formulas as answers, model industrial mathematical problems, automated-proven properties and various other manipulations.

Symbolic processing is a special feature of these systems. That is the excitation signals may represent not only numbers, even the symbols, and it is analyzed what happens to them during processing. The parameters of the system can be analyzed subsequently appointing of additional requirements on the desired response, so, on this way the analysis is used in system design. Symbolic processing is concerned with alternative forms, partially-specified domains, and symbolic derivations in general and it can be treated as a transformation of expression trees, such as symbols for operations, variables, constants, and simplification, and plenty expressions, functions and equations can be derived in closed form using symbolic processing, as well as the error-free derived simulation code can be obtained.

Plenty instances of expert-based analysis will be shown: design of high-speed digital filters, simulations of transceiver OFDM (Orthogonal Frequency

Division Multiplexing), expectation-maximization algorithm in modeling of propagation channels, linear systems, non-linear systems, modeling, and simulation. All of these general explanations are illustrated on practical applications in symbolic close-form. The software packet Mathematica is used to demonstrate all mentioned examples.

This chapter is organized as follow: second section describes a general statement of Computer Algebra, in the third section fundaments of symbolic processing are explained. Fourth section provides examples of using the computer algebra and symbolic processing in detail. Last section gives the significance of the method and future applications.

II. General Statement of Computer Algebra and Symbolic Processing

The first computer algebra programs appeared more in 1990. year, but most of them did not have expansion as numerical-only systems. The main reason was that they were able to solve simple practical problems, the calculation time was very long and often the problem of presentation of results was present primarily because of the limited storage space. Computer algebra systems work with symbols, and numbers are just a narrower group of symbols that perform some operations, such as addition and multiplication operations. Therefore, writing programs in computer algebra systems can be seen as a set of instructions that manipulate symbols and can be used to perform a much wider range of works [1]. One of the simplest examples of the use of computer algebra is addition of two numbers. For a short example the addition of one positive and one negative number will be illustrated, i.e. subtraction operation.

Let us suppose that the numerical values a=0.2 and b=1-0.8 are given. We expect that the difference between the numbers a and b is zero, but using the programs Matlab and Mathematica we get a-b = 5.5511×10-17. The reason for this surprising result is easy to find if we know how numerical computing systems represent these two numbers. For example, the Matlab command sym (a, 'd') shows that the number is presented as the number 0.20000000000000001110223024625157 not as 0.20000000000000. In the binary system the number of 0.2 can not be represented exactly with a finite number of bits. Although the Matlab command window shows that b=0.20000000000000, the command sym (b, 'd') gives

0.199999999999999995559107901499374 and this is the number used in the calculations. Contrary to our expectation that the numbers a and b are the same as those presented in different ways as well as the difference of these two numbers are different than 0. Although the difference of a and b is not equal to zero, the command == in Mathematica, which tests whether the left and right side of the expression a == b equal, evaluates to True, i.e. it shows that the numbers a and b are equal. The same command in Matlab evaluates to 0, it shows that the numbers a and b are different. Seemingly the same command, which symbolically represents the double equals sign ==, gives different results in the two math programs. Below, the practical implementation code of both of these examples will be shown. The analyzed example shows that numerical systems perform operations with numbers that differ from those shown, and generally we do not know exactly which the numbers are used by programs. For example, in Matlab, the command sign(sin(2*pi)) gives as result a "-1", and a command sign(sin(3*pi)) gives a "+1", although in both cases the correct value equal to 0. The computer algebra systems work with the exact values. For example, the command simplify(sym('sin(2*pi)')) and simplify(sym('sin(3*pi)')) give the same exact value of 0, so the sign(0)=0. Number can be represented as a rational number; assignment of the number can be done symbolically a=sym(1/5). The b=sym(1-4/5) and then both numbers have the same value a=1/5, b=1/5 in both cases, the numbers pi=π and 1/5 are approximate values for symbols that contain the correct values. For example, the a=sym(sin(pi/4)) gives the correct result a=sqrt(1/2), in contrast to the numerical calculation that gives results accurate to approximately 15 decimal digits sin (pi/4)=0.70710678118655. Also, one of the simplest examples of the difference between numerical and computer algebra system is solving of integration of sinusoidal signal form. Namely, the numerical code calculates numerical values with a certain step of independent variables, and then it finds signal corresponding to the integral of the sinusoidal signal by simply adding up. It can not be seen to get cosine signals using by integration of sine signals with a small number of numerical values, and a great number of numerical values can lead to significant errors due to the accumulation of the representation of numbers with a finite number of digits. In the case of using the computer algebra system, the initial waveform is a symbol of the sinus, an operation that is performed is represented as symbol of integral, and result of processing is a symbol of cosines. Regardless of the range of the independent variables where the signal is observed, result of processing always is a symbol that is instantly recognized and which is not subject to error processing. For more complexes of systems and signals, the possibility of errors with numerical computer systems is greater, so efficiency of computer algebra system becomes more important. The characteristics of signals, which are represented by symbols as well as and attributes of a system that are defined by the symbolic operations, can be defined in the computer algebra systems. If the signals are presented by numbers, and operations systems are presented by the set of operations of addition and multiplication, then the computer algebra system will work the same as numerical. However, if the analyzed system is presented by system of difference equations and the excitation signals are represented as a function of a variable, then results of process can also be presented as functions. This means that the computer algebra systems can transform the signal presented in one form to another form in the same way as in the classical mathematical theorem or analyzes properties of functions. The computer is able to automatically perform the necessary symbolic operations until the desired shape instead of ourselves on the same way as we derive by hand as well as proof of a theorem. It shows that the computer algebra systems look like some kind of intellectual amplifier, or they have the ability to increase our capacity of thinking and reasoning, and that, for example, in a split second they find the integral of a complex function just like a calculator multiplies two numbers represented by a large number of digits. One should be careful, because as there are no algebraic solutions for all integrals, computer algebra systems can not solve all the problems. Therefore, the program is reduced to writing a program for communication between users and computers that the computer determines that the accumulated knowledge should be used, as well as when and how to apply that knowledge to solve practical problems.

In this section, the efficiency of computer algebra system will be demonstrated through simple practical examples. For example, a block diagram of the analyzed system can be drawn using a palette of graphical symbols of discrete system, such as multiple input and output ports, that the parameters or symbolic values would be defined using by block certain, and then to automatically get all the transfer functions of all inputs to all outputs. The response in the time domain for arbitrary excitation signals can be determined from the transfer function. In some complex cases, such as multi-stage filters for high-speed processing, transfer functions are difficult to calculate the classical methods, while using the computer algebra system in a short time can obtain accurate solutions. The results can be used for further analysis of these filters such as, for example, evidence of complementarily by force. In a similar way we can compare different systems and to show what is different, or prove that they have the same functionality.

Computer algebra systems can be used to simplify rational functions, factor polynomials, find the solutions to a system of equation, give general formulas as answers, model industrial mathematical problems, and various other manipulations. Computer algebra deals with arithmetic on defined algebraic structures, polynomials, matrices, algebraic functions, and may involve symbols parameters and indeterminate. It is based on well-defined semantics [1], compose constructions, and algebraic algorithms and it is concerning with alternative forms (factored, expanded), partially-specified domains, and symbolic derivations in general. Application of computer algebra system allows to devote to selected problem, such as how to solve some problem on way that the non-creative work of finding solutions leave to the computer, such as for example, solving systems of equations or integrals of complex functions. These systems should serve as machines that have the ability of mathematical thinking and reasoning in order to liberate us uncreative work and not to deal with the mathematical methods. Although at first glance it seems that the use of computer algebra system is very complicated, and it is also one of the reasons that are not as popular as numerical systems, most users will often use specialized packages or only a small number of symbolic functions. Specialized programs intended for use computer algebra systems have a graphical user interface so that certain operations are performed simply by activating an icon in the same way as standard programs used for drawing or word processing. Each command has a corresponding auxiliary text with short instructions for use and comprehensive guide for those who want to use some specific features of the symbolic function.

Symbolic simulation is a formal technique to calculate the behavior or a characteristic of a system (e.g. digital system, electronic circuit, or continuous-time system) with an independent variable (sample index, time, or frequency), the dependant variables (sample values, signals, voltages, and currents), and (some or all) the system elements represented by symbols. The symbolic technique is complementary to numerical techniques (where the variables and the elements are represented by numbers) and qualitative analysis (where only qualitative values are used for signals, such as increase, decrease or no change) [3]. A symbolic simulator is a computer program that receives the system description as input and can automatically carry out the symbolic analysis and thus generate the symbolic expression for the desired system characteristic. Majority of the symbolic simulation research has concerned the analysis of linear systems in the frequency domain. For lumped, linear time-invariant (LTI) systems, the symbolic transfer functions obtained are rational functions in the complex frequency variable (s for continuous time systems and z for discrete-time systems) and the system elements that are represented by a symbol (instead of numerical value). Depending on whether the complex frequency is the only variable, and whether all of the system elements are characterized by symbolic parameters, we have three levels of symbolic representation: rational transfer function of the complex frequency with numerical coefficients, partially symbolic transfer function, and fully symbolic transfer function [4]. Majority of reported symbolic simulators were written as application packages (toolboxes) for computer algebra systems (CAS), such as Mathematica [5]. Implementing a symbolic simulator as a computer algebra system toolbox has many advantages: built in functions for the basic and advanced symbolic computations are highly optimized within the computer algebra system, so the simulator relies on the most efficient code to be used for the analysis kernel, the rich functionality of the computer algebra system can be used for post processing of the results generated by the simulator, the supreme graphics/multimedia computer algebra system support can be exploited for visualization, typesetting, or animation of the simulation results.

General Application of Symbolic Simulation

The value of symbolic analysis is well recognized in both industry and academia. In industry it has been used as an aid in the design of systems and circuits. In academic institutions it has been found useful as an instructional aid. There are many reasons why one may be interested in symbolic simulation. A few of the more important ones are as follows: frequency response calculation, parameter iteration, sensitivity analysis, insight and instructional aid. For Frequency response calculation we suppose that an accurate magnitude response curve is desired over a frequency range with a specified frequency step. With a numerical program such as Matlab Simulink [6], the same network or system will have to be analyzed many times. On the other hand, if we obtain the symbolic transfer function first as a rational function with real coefficients, then we need only evaluate it at different values of frequency – obviously a much simpler task. When we talk about parameter iteration, for solving piecewise linear resistive networks the exhaustive segmentation combination method is conceptually simple and can produce all solutions. In this case, the use of a symbolic technique is a natural choice. If we first obtain symbolic expressions for the response, then it is only necessary to substitute the parameters for each segment combination into the expressions and check whether the solutions lie within the ranges defining the segments. If not, that particular segment combination does not yield a solution and we also proceed to another segment combination. Sensitivity analysis is used in the design of any system, it is important to know the effect on the network performance due to the variation of some element values. A precise measure of the effect can be expressed in terms of the sensitivity functions. These functions contain partial derivatives of the system response with respect to element values (system parameters), so symbolic computation becomes a natural approach: find symbolic system function, compute (symbolically) required derivatives, and determine (a closed-form expression) for the desired sensitivity function. Filter design by optimization. In this approach of filter design, a reasonable network configuration is first proposed, and initial element values are selected from an approximate analysis. The actual frequency response is then calculated and compared with the specified response. The process is repeated until the error is minimized. It is seen that in this approach of network design, the response of the network has to be calculated at many frequency points, and at many different sets of element values. Obviously, if a symbolic transfer function can be obtained first, repeated evaluation of the transfer function will be much simpler job than repeated analysis of the network. Talking about insight we talk that symbolic transfer function can provide better insight than numerical solutions. By inspection of the symbolic transfer function, it might be immediately clear how a parameter (or an element value) contributes to the performance and behavior of the system. Without a symbolic transfer function, these conclusions could only be reached after the analysis of many numerical cases, and even then some degree of uncertainty still exists [7], [8], [9], [10], [11], [12], [13], [14]. Finally, instructional aid is used for beginning linear circuit courses and signals and systems courses contain many exercises that ask students to derive the expressions for the transfer function, input impedance, voltage gain, current gain, etc. It is very easy for students (in fact, even for the instructors!) to make minor "math errors" that lead to incorrect answers. Therefore, it would be very helpful to the students to have a tool that (symbolically) checks their answers for correctness. Discovery of a wrong answer before handing in their work enables them to redo the problem and make corrections. Symbolic analysis even has the potential to improve the training of young analog circuit designers and to guide more experienced through second-order phenomena such as distortion [8].

Modern research in the field of engineering science usually starts with theoretical analysis. The next step can be simulation in order to prove the theoretical assumptions. The final step is practical implementation and measurements. Usually, the theoretical derivations are made on ideal parameter values, such in infinite number of samples or known signal values for time from $-\infty$ to $+\infty$. Simulation is used for simulation of systems in order to gain insight into their functioning. For a finite number of measured values, one can expect that the result agrees well with the simulation [15].

Symbolic processing can help to error-free derived simulation code, and to find typewriting errors in published results. Next, symbolic processing can be used for finding the processing errors as the closed-form expression for computing the number of required iteration steps; or the error function due to finite word-length [14, 16]. Therefore, symbolic processing can help gain insight into how a system works, which is preferred to experimenting with numerical simulations.

Symbolic processing plays an important role in education. Even more it can be used in practice because the numerical tools may have problems with accuracy and a significant increase in the complexity. Primarily, the analysis area is essentially continuous and infinite [17]. Symbolic processing repairs certain disagreement between theoretical performance and numerical simulations. Numerical processing may show unsatisfactory results [18] and it can drive the researchers to erroneous conclusions. Using the symbolic tools, it is easy to find and prove mistakes and gain a better insight into the whole process of analysis, simulation and modeling. Symbolic processing works with symbols. Any symbol can be replaced by a number when symbolic expressions become impractical, for example, for plotting a response to a specific excitation. Therefore, writing programs using symbolic processing can be seen as a set of instructions that manipulate the symbols and can be used to perform a much wider range of activities [19]. The efficiency of symbolic processing becomes more important if systems and signals are more complex. There is no algebraic solution for all forms of processing, and symbolic processing will not be able to solve all problems. Even then, the final result can be expressed in a form of special function that can be computed for known numerical values of system parameters. Finite wordlength effects play one of significant roles in symbolic processing [14]. For example quantized filter coefficients make the transfer function different from the ideal one (with infinite-precision coefficients). Those errors can be efficiently predicted by the coefficient sensitivity analysis that is based on symbolic derivation and substitution operations. On the contrary, the computation of the variance of the uncorrelated noise due to rounding the output of the multiplier element is very non-linear and the numerical analysis was extensively used in practice. The computer algebra system is capable to successively solve this non-linear problem in a straightforward manner. SchematicSolver is package of Mathematica which helps to solve, simulate, analyze many processes using GUI (Graphic User Interface), and finally, generates implementation code. For this case the SchematicSolver is used to build a symbolic object of the system as schematic specification and to identify the symbolic system parameters. After that, Mathematica can be used to analyze the symbolic object, to derive the partial transfer function that corresponds to the identified system parameter, to express the noise variance in terms of the pole magnitude and the pole argument, to generate a noise transfer function, and finally, to derive the variance of uncorrelated noise due to rounding.

Original Design Procedure

The design of a system starts with a set of requirements and it usually consists of approximation, realization, study of imperfections and implementation [14]. The production cost may depend on the type of components, packaging, methods of manufacturing, and testing. If the requirements are not met, or the cost is very high, the realization step or the approximation step must be redone, and then, the development cost can become very large. The computer algebra systems provide a new original design procedure. Starting with the preferred efficient implementation, the functional description (say the transfer function) can be expressed in terms of symbolic parameters. Thus, approximation is no more the first step, but it becomes the last step in the design of the system. The symbolic design parameters are b, k0, k1, k2, and k3. Assume the requirements which imply that H1 should be zero for z→1

num3=Numerator[H1//Together]/.z→1//Factor

$$(1 + b)^5 \, k0 \, (1 - k1 - k2 - k3 - k1\,k2 - k1\,k3 - k2\,k3 + k1\,k2\,k3) == 0.$$

The coefficient k0 has been already computed in the previous section. The poles of a stable discrete system should be within the unit circle, that is, $1<b<1$. Finally, k2 can be expressed in terms of k1 and k3

Solve[num3 == 0, k2]//Flatten//FullSimplify

$$\left\{\mathrm{k2} \to \frac{-1 + \mathrm{k1} + \mathrm{k3} + \mathrm{k1k3}}{-1 + \mathrm{k1}(-1 + \mathrm{k3}) - \mathrm{k3}}\right\}$$

The two expressions for k0 and k2 are derived in terms of k1 and k3, directly from the schematic, using the two properties of the high-speed filter [20]. No assumption on the filter approximation function has been made. Furthermore, the closed-form expressions show that k0 and k2, and the corresponding properties, do not depend on the parameter b. This means that the coefficient b can be independently optimized to satisfy other requirements. Other requirements can be met by a simpler two-variable numerical optimization (k1 and k3), instead of a much more complex optimization of the five design parameters. This way, computer algebra system helps to come up with an efficient implementation and balance time spent for optimizing the design cost against completing the project quickly.

PRESENTATION OF NUMBERS IN NUMERICAL AND COMPUTER ALGEBRA SYSTEMS

In this section of text, we going to display how examples are demonstrated on practical way described in the previous section [1]. Mathematica is the best computer tool for symbolic computation. The basis of the record is expression, and therefore symbolic record is the basic structure for the representation of variables. The first input cell to define a variable which is assigned the value of b, Figure 1.

```
In[1]:= a = b

Out[1]= b
```

Figure 1. Direct assignment of symbolic value.

The execution of this assignment would be reported an error using this numerical tools if numerical value for b not previously known. In

Mathematica there was no error because the variable is assigned a value of the variable `b`, and whatever it was, including that value of `b` is not known. Therefore, the value `b` appears in the output cell which shows the result of the execution of the input cell.

```
In[2]:= c = a + 2

Out[2]= 2 + b
```

Figure 2. Calculations of symbolic values.

The result will appear in the symbolic value of `b` if the variable `a` would be used in next input cell, Figure 2. The Mathematica can be used for numerical calculations. If we repeat the example analyzed in Matlab, result of several successive calculation will be obtained without the expected value such as 0, so the school children would know to do it, Figure 3.

```
In[3]:= d = ((((1 - 0.2) - 0.2) - 0.2) - 0.2) - 0.2

Out[3]= 5.55112 × 10^-17
```

Figure 3. Numerical calculations can give unexpected results.

Of course, the same example can be repeated in a different way, using fractions, Figure 4. The result is exactly and has a value of 0.

In[4]:= $e = \left(\left(\left(\left(1 - \frac{2}{10}\right) - \frac{2}{10}\right) - \frac{2}{10}\right) - \frac{2}{10}\right) - \frac{2}{10}$

Out[4]= 0

Figure 4. Symbolic computation gives the correct result.

The simplest way to learn the basics of working in the software Mathematica is to first read "*the First Steps with Mathematica*". This chapter gives only the most important features. One of the most important functions is `Simplify`, example in Matlab is given in Figure 5. Here, the term simplification implies that the expression presents least members from all the possible combinations.

In[5]:= Simplify[Sin[x]^2 + Cos[x]^2]

Out[5]= 1

Figure 5. Simplifying expressions.

It is important to note that the arguments of the function are enclosed in square brackets, not parentheses that are commonly used. The reason is that parentheses are used for grouping of members. For instance the system of equations or quadratic equations can be solved, Figure 6. Input cell is written in 1D form, but the result is displayed in 2D form. Since there is more than one solution, all solutions are given as a set in curly brackets, each of the solutions is also a set of expressions.

```
In[6]:= equation = α x^2 + β x + γ
        s = Solve[equation == 0, x]
```

Out[6]= $x^2 \alpha + x \beta + \gamma$

Out[7]= $\left\{\left\{x \to \frac{-\beta - \sqrt{\beta^2 - 4 \alpha \gamma}}{2 \alpha}\right\}, \left\{x \to \frac{-\beta + \sqrt{\beta^2 - 4 \alpha \gamma}}{2 \alpha}\right\}\right\}$

Figure 6. Numerical calculations can give unexpected results.

It should be noted that the input cell uses two equal sign, ==, which indicates about the equation and not giving value. There is no solution for the allocation of an equal sign, =, but has an arrow →. The arrow means that the decision rules that may apply. Member of the array can be obtained by using double square brackets [[]], Figure 7. The first element of the array is also an array, so it is in parentheses. The letters of arbitrary type can be used in mathematics, in this example the Greek letters are used.

Functions and Domain of Variables

When we have rules, such as solutions which are indicated by an arrow, these rules can be used in other expressions. For example, if we look at Figure 7, the solution of the equation is used by replacing the term from which we started, and the expression is equal to 0, which is the expected value.

```
In[8]:= s[[1]] // Simplify
```

$$\text{Out[8]= } \left\{x \to -\frac{\beta + \sqrt{\beta^2 - 4\,\alpha\,\gamma}}{2\,\alpha}\right\}$$

```
In[9]:= equation /. s[[1]] // Simplify

Out[9]= 0

In[10]:= equation /. s[[2]] // Simplify

Out[10]=
         0
```

Figure 7. Simplifying expressions.

The rule means that the expression on the left side in front of the sign `/.`, i.e. the variable that we have called expression, the `x` is replaced by the expression that first, i.e. the second member of the series of s. Double slash means that the expression is sampled after the introduction of substitution in expression. Mathematics has a rich set of algorithms for solving systems of equations and inequalities, including the definition of domain in which there is a solution. In case there is no solution in symbolic form, the solution can always be obtained numerically, of course, if solution exists. The Mathematica can solve systems of difference equations. The concept of Mathematica is working with notebooks. One notebook contains everything needed for whole document to be entirety, including the document title and chapter, the cells with text descriptions, codes, the results of executing of the commands and functions as well as visualizing of the results.

Since everything is in one document, subroutines must be defined within the notebook. It is therefore adopted the following notation for defining functions, Figure 8.

```
In[1]:= y[a_, x_] := a + x^2

In[2]:= y[1]

Out[2]= y[1]

In[3]:= y[2, e^z]

Out[3]= 2 + e^(2 z)
```

Figure 8. Defining and calling of function.

The function y is defined in the input cell 1 which takes two arguments, `a` and `x`, such that `a` and `x` can be any numerical value, symbol or expression. We can get the same expression, as in the input cell 2, if we want to perform a function with one argument because there is no defined function with one argument. The function with two arguments is performed in the cell 3, and the number 2 appears on the first place, and the expression is on the second place. In this example, the setting of the argument is used with `a_`, where the bottom dash behind of symbols means that any expression can be used instead of the symbol. The colon before the equality means that the `:=` defines functions, and execution of it happens after calling of this function. The following is an interesting example, Figure 9.

```
In[4]:= w = a^2 + b

Out[4]= a^2 + b

In[5]:= q = w /. Plus → Power

Out[5]= (a^2)^b
```

Figure 9. Changing of functionality substituting of an operation by another.

At first glance, there is no reason to get different expression after the introduction of rules replacing the `Power` with `Plus`. The same expression can be displayed in three ways as on Figure 10.

```
In[6]:= a^2 + b

Out[6]= a^2 + b

In[7]:= Plus[b, Power[a, 2]]

Out[7]= a^2 + b
```

Figure 10. Various previews and structures of records.

This means that all operations are encoded as functions, including basic ones such as, for example, the addition or exponent. These types of substitutes allow complex manipulation of expressions and represent the software Mathematica as very powerful tool for the manipulation of expressions. When Mathematica does not know how to calculate a given task set in the input cell, then the cell in the output appears the same expression, as an example of a cell 9, Figure 11. The reason that does not appear in the result is that there is no information about the domain of x.

```
In[9]:= Simplify[√(x^2)]

Out[9]= √(x^2)

In[10]:= Simplify[√(x^2), x < 0]

Out[10]=
   -x
```

Figure 11. The result depends on the domain variables.

If we define the domain variable as the second argument to the simplification, for example, that the variable is real and negative, then simplification can be done and the result can be obtained. The result may be different than expected, as on Figure 12, and the difference is consequence of different view of domains. In fact, when someone says to calculate the third root of -1, it is usually thought that the result is real, and most will answer that it is -1. However, the majority of computer tools include the solutions in the domain of numerical solving of complex numbers and the result is a number that has the smallest angle with respect to the real axis, which is given in a cell 12. The output cell 11 contains essentially the same record as the input cell, but a symbolic record does not present clearly how it would be used the result later if numerical computation is used. Although there are three solutions, as it is shown in the cell 13, the first solution would be used if numerical processing is used, and the secondly is the one most used to work without a PC.

In[11]:= $u = \sqrt[3]{-1}$

Out[11]= $(-1)^{1/3}$

In[12]:= u // N

Out[12]= 0.5 + 0.866025 i

In[13]:= $u3 = \left\{\frac{1}{2} + i\frac{\sqrt{3}}{2}, -1, \frac{1}{2} - i\frac{\sqrt{3}}{2}\right\}^3$ // Simplify

Out[13]= {-1, -1, -1}

Figure 12. The result that differs from the expected solution depending on the domain.

Other features of Mathematica will be shown in the examples in the following sections. As in the case of Matlab, additional knowledge that is in the toolboxes and application packages can be loaded and used with the built-in Mathematica functions.

Time-Domain Analysis of Discrete Systems

In symbolic processing, the signal can be represented as a formula, rather than as a sequence of numbers, but in symbolic close-form. Thus, the value of a signal might be known in terms of a formula. In a similar manner, signal-processing operators, which correspond to the building blocks of systems, are maintained in symbolic form. This enables computer algebra system to simplify, rearrange and rewrite symbolic expressions until they take a desired form. When one of the operators is applied to a function, no evaluation takes place. The resulting function is stored in the symbolic form until it becomes convenient to compute it explicitly. Consider a discrete-time recursive system whose transfer function is known, say the numerator is $(1 + 2z^{-1}+z^{-2})$ and the denominator is $(1+0.5z^{-2})$. First, the schematic of the system is created by the command

```
{schematic,{inpCoord},{outCoord}}=
TransposedDirectForm2IIRFilterSchematic[{{1, 2, 1}, {0, 1/2}}];
```

Next, the input element and the output element are added:

```
system=
Join[schematic, {{"Input", inpCoord, X},{"Output", outCoord, Y}}];
```

The block-diagram of the system is shown in Figure 13. Transfer function can be computed directly from the schematic system as follows:

```
{tfMatrix, systemInp, systemOut}=
DiscreteSystemTransferFunction[system];
tf=tfMatrix[[1, 1]];
```

Suppose that the input signal can be represented by a formula
sineSignal=Sin[n/5];

Traditionally, the output signal can be found by (1) multiplying the z-transform of the input signal by the
transfer function and (2) taking the inverse z-transform of the product

```
sineTransform=ZTransform[sineSignal, n, z];
response=InverseZTransform[sineTransform*tf, z, n];
```

which yields

$$\frac{8\,\mathrm{Cos}[1/10]^2(2^{-n/2}\mathrm{Cos}[n\pi/2]\mathrm{Sin}[1/5]-2\mathrm{Sin}[(1-n)/5]+\mathrm{Sin}[(1+n)/5])+2^{1/2-n/2}(\mathrm{Sin}[1/5]-4\mathrm{Sin}[2/5])\mathrm{Sin}[n\pi/2]}{5+4\mathrm{Cos}[2/5]}$$

In many cases, the traditional approach to computing the response of a discrete system, for known input signal, cannot be accomplished successively.

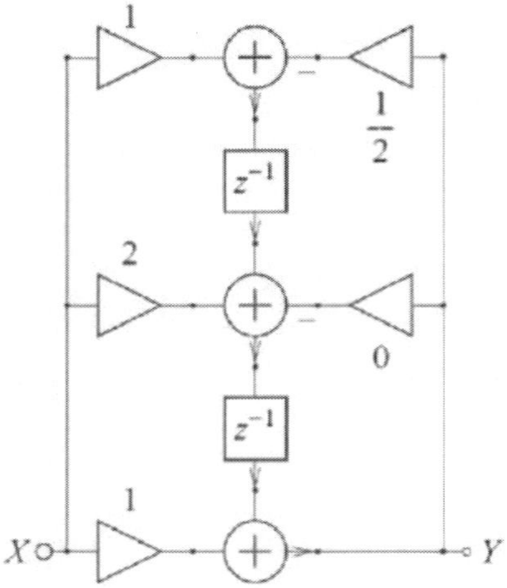

Figure 13. Block-diagram of the recursive system that has been automatically generated for given parameters.

Even a very powerful computer algebra system can fail to find the z-transform or the inverse z-transform. Symbolic processing can be used instead to overcome the problem. For the schematic of Figure 13, symbolic processing can be carried out as follows: automatically generate a code that implements the system, compute the input sequence whose elements can be symbols, numbers, or formulas, and process the input sequence with the code

```
DiscreteSystemImplementation[system, "imp"];
inSeq=UnitSineSequence[8, 1/(10 π), 0];
{outSeq,finals}=
DiscreteSystemImplementationProcessing[inSeq, {0, 0}, {}, imp];
```

For example, the seventh element of the output sequence is not a number, it is an expression

$$\frac{1}{4}\left(2\mathrm{Sin}\left[\frac{1}{5}\right]-\mathrm{Sin}\left[\frac{2}{5}\right]-4\mathrm{Sin}\left[\frac{3}{5}\right]+2\mathrm{Sin}\left[\frac{4}{5}\right]+8\mathrm{Sin}[1]\right)$$
$$+\mathrm{Sin}\left[\frac{6}{5}\right].$$

The expression derived by using the z-transform may look different, such as

$$\frac{8\mathrm{Cos}[1/10]^2(-1/8\mathrm{Sin}[1/5]+2\mathrm{Sin}[1]+\mathrm{Sin}[7/5])}{5+4\mathrm{Cos}[2/5]}.$$

Simplifying the difference of these two expressions shows that the output sequences in both cases are the same. Symbolic processing of a symbolic sequence, by a system that can be described with symbolic parameters, always returns a symbolic output sequence, even in the cases when computer algebra system fails to compute the z-transform or the inverse z-transform. In addition, at any step of the processing, any symbol can be replaced by a number.

III. Applications

There are many areas where computer algebra and symbolic processing found their place, but we will focus in this section solely by how those are used in field of telecommunications. The text below will illustrate and describe in details the examples of the use of these methods with a strong emphasis on the benefits compared to existing approaches. The first example shows the typical algebraic loop that numerical tools can not display solution, but using the tools of symbolic solution it is obtained. Second example is application of computer algebra and symbolic processing in the design of high-speed filter. The third example shows a model post-processing with advanced symbolic simulation and design. The fourth example shows in detail the simulation and analysis of OFDM Transmitter, and last one shows the benefits of symbolic analysis as universal tool for deriving properties in expectation-maximization algorithm.

Example 1 - Symbolic Analysis of Algebraic Loop

Symbolic analysis of systems is inherently immune to the problem imposed by algebraic, or implicit, loops occurring when two or more blocks with direct feed through of their inputs form a feedback loop [14]. On the other hand, in numerical simulations, i.e. in the Matlab/Simulink interpreter, algebraic loops considerably reduce the speed of a simulation and may be unsolvable. Consider a simple feedback system and the corresponding model in Matlab/Simulink as shown in Figure 14. After running the simulation, an error message appears: ''...*Stopping simulation. There may be a singularity in the solution* ...'. The same system can be symbolically solved with Mathematica and SchematicSolver as shown in Figure 15. The computer algebra system successfully and accurately computes the required response. It is important to point out that the schematic specification (denoted as schematic in this example) is a symbolic object that can be used for both visualization and solving.

MATLAB-Based Approach

Matlab is supplemented by the Symbolic Math Toolboxes which incorporate symbolic computation into the numerical environment. The computational engine underlying the toolboxes is the kernel of Maple. Several application toolboxes utilize the Matlab symbolic features.

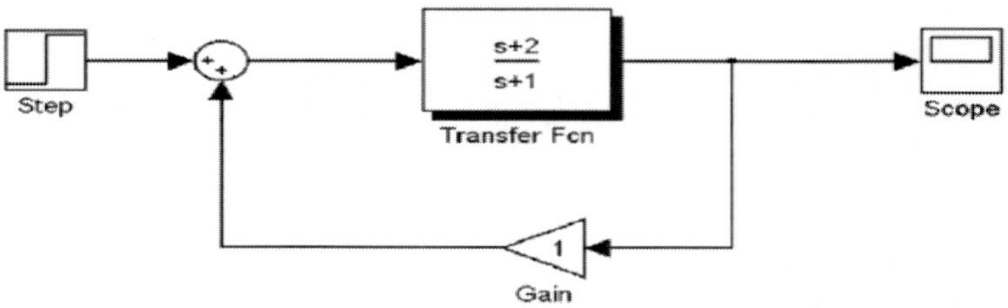

Figure 14. Simulink model of a system with an algebraic loop.

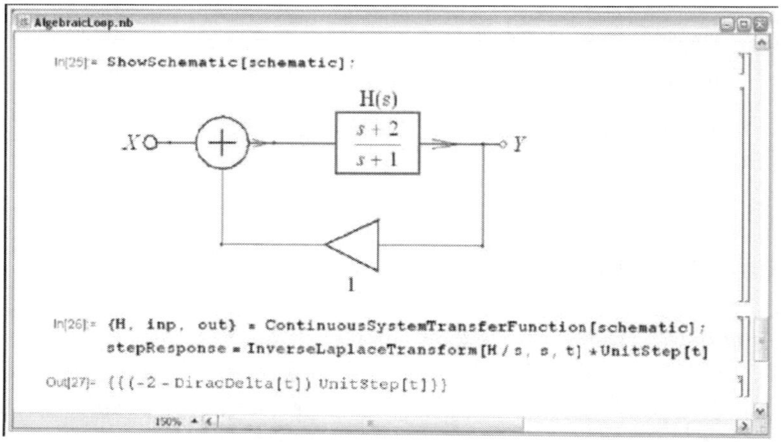

Figure 15. SchematicSolver model of a system with an algebraic loop.

A closed loop symbolic representation of the control system is generated automatically using symbolic computation. The parameter space approach and various derivatives can be applied to this parametric model.

Example 2 - High-Speed Filter Design Using Mathematica

In this example, we focus on the application of computer algebra system to high-speed filter design [21], [22] and demonstrate the use of symbolic computation [17]. Our computations were performed with Mathematica [5], and the application package SchematicSolver [23].

Schematic Representation of Systems

Traditional mathematical formulas, that describe a filter or a system, can give relations and results, but typically cannot capture the processes of calculation that lead to them. System models are often best represented directly as "visualized" algorithms by means of pictorial block-diagram representations – schematics.

SchematicSolver is an application package in which we can draw block-diagrams that represent various systems. The graphical representation is essential for supporting a designer's view of implementation. SchematicSolver palettes provides a point-and-click graphical user interface for building models as block-diagrams, using mouse operations for performing drawing tasks. We can draw the models just as we would with pencil and paper.

Here is a portion of the list called hsSystem:

```
hsSystem={
{"Multiplier", {{6, 0}, {6, 3}}, k1},
{"Delay", {{4, 5}, {4, 7}}, 1},
{"Adder",{{7,8},{8,5},{9,8},{8,9}},{1,1,2,0}},
{"Input", {0, 8}, "X"},
{"Output", {9, 8}, "Y1L"},
...,
{"Line", {{6, 8}, {7, 8}}} }
```

We can easily build new models from these automatically generated schematics. For example, given the number of stages of a high-speed filter, numStages, we can automatically generate a list of filter parameters parameterSymbols that consists of the list of the symbolic coefficients of the FIR filter, p, and the symbolic coefficient of the halfband IIR filter, b:

```
numStages=3
p=UnitSymbolicSequence[numStages+1, k, 0]
parameterSymbols=Join[{b}, p] // Flatten
{hsSystem,inpCoordsHS,outCoordsHS}=
HighSpeedIIR3FIRHalfbandFilterSchematic[parameterSymbols,{0,0}];
```

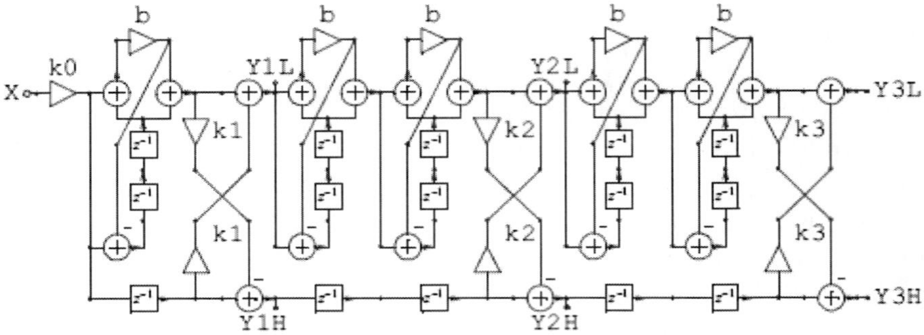

Figure 16. Schematic of the high-speed filter.

We can automatically add the input `X` and the output `Y3L` to the schematic `hsSystem` using the coordinates `inpCoordsHS` and `outCoordsHS`. In addition, we can add other elements, such as output `Y1L`, by clicking on the `Output` element in the palette and clicking when the mouse position is over the required coordinate in the drawing workspace. The schematic can be drawn using the command `ShowSchematic[hsSystem]`.

We can draw a schematic of a system and then compute the system transfer function directly from the schematic by clicking the button Solve on the palette. Alternatively, we can issue the command

```
{tfMatrix, systemInp, systemOut}=
DiscreteSystemTransferFunction[hsSystem];
```

which finds the transfer function matrix of this single-input six-output MIMO system. The partial transfer function from the input X to the output Y1H is

```
h1H = tfMatrix[[2,1]]// FullSimplify // Together
DiscreteSystemDisplayForm[%]
```

Derivation of h3H=Y3H/X=tfMatrix[[6,1]] by hand will be very time-consuming and error prone:

$$
\begin{aligned}
-\frac{1}{z^5\,(b+z^2)^5}
&(k0\,(-b^5 + b^4\,k1\,z - 5\,b^4\,z^2 + b^3\,k1\,k2\,z^2 + b^3\,k2\,k3\,z^2 + \\
&4\,b^3\,k1\,z^3 + b^5\,k1\,z^3 + b^2\,k2\,z^3 - b^2\,k1\,k2\,k3\,z^3 - \\
&10\,b^3\,z^4 + 3\,b^2\,k1\,k2\,z^4 + 2\,b^4\,k1\,k2\,z^4 + b\,k1\,k3\,z^4 + \\
&3\,b^2\,k2\,k3\,z^4 + 2\,b^4\,k2\,k3\,z^4 + 6\,b^2\,k1\,z^5 + 4\,b^4\,k1\,z^5 + \\
&2\,b\,k2\,z^5 + 3\,b^3\,k2\,z^5 + k3\,z^5 - 2\,b\,k1\,k2\,k3\,z^5 - \\
&3\,b^3\,k1\,k2\,k3\,z^5 - 10\,b^2\,z^6 + 3\,b\,k1\,k2\,z^6 + 6\,b^3\,k1\,k2\,z^6 + \\
&b^5\,k1\,k2\,z^6 + k1\,k3\,z^6 + 4\,b^2\,k1\,k3\,z^6 + 3\,b\,k2\,k3\,z^6 + \\
&6\,b^3\,k2\,k3\,z^6 + b^5\,k2\,k3\,z^6 + 4\,b\,k1\,z^7 + 6\,b^3\,k1\,z^7 + \\
&k2\,z^7 + 6\,b^2\,k2\,z^7 + 3\,b^4\,k2\,z^7 + 5\,b\,k3\,z^7 - k1\,k2\,k3\,z^7 - \\
&6\,b^2\,k1\,k2\,k3\,z^7 - 3\,b^4\,k1\,k2\,k3\,z^7 - 5\,b\,z^8 + k1\,k2\,z^8 + \\
&6\,b^2\,k1\,k2\,z^8 + 3\,b^4\,k1\,k2\,z^8 + 4\,b\,k1\,k3\,z^8 + 6\,b^3\,k1\,k3\,z^8 + \\
&k2\,k3\,z^8 + 6\,b^2\,k2\,k3\,z^8 + 3\,b^4\,k2\,k3\,z^8 + k1\,z^9 + \\
&4\,b^2\,k1\,z^9 + 3\,b\,k2\,z^9 + 6\,b^3\,k2\,z^9 + b^5\,k2\,z^9 + 10\,b^2\,k3\,z^9 - \\
&3\,b\,k1\,k2\,k3\,z^9 - 6\,b^3\,k1\,k2\,k3\,z^9 - b^5\,k1\,k2\,k3\,z^9 - \\
&z^{10} + 2\,b\,k1\,k2\,z^{10} + 3\,b^3\,k1\,k2\,z^{10} + 6\,b^2\,k1\,k3\,z^{10} + \\
&4\,b^4\,k1\,k3\,z^{10} + 2\,b\,k2\,k3\,z^{10} + 3\,b^3\,k2\,k3\,z^{10} + \\
&b\,k1\,z^{11} + 3\,b^2\,k2\,z^{11} + 2\,b^4\,k2\,z^{11} + 10\,b^3\,k3\,z^{11} - \\
&3\,b^2\,k1\,k2\,k3\,z^{11} - 2\,b^4\,k1\,k2\,k3\,z^{11} + b^2\,k1\,k2\,z^{12} + \\
&4\,b^3\,k1\,k3\,z^{12} + b^5\,k1\,k3\,z^{12} + b^2\,k2\,k3\,z^{12} + b^3\,k2\,z^{13} + \\
&5\,b^4\,k3\,z^{13} - b^3\,k1\,k2\,k3\,z^{13} + b^4\,k1\,k3\,z^{14} + b^5\,k3\,z^{15}))
\end{aligned}
$$

Note that we have to specify only one numerical value – the number of stages of the high-speed filter.

Deriving Filter Properties

We can demonstrate the power of computer algebra system by proving the power-complementary property of the high-speed filter:

$$H_{3L}(z)H_{3L}(z^{-1})+H_{3H}(z)H_{3H}(z^{-1})=1$$

Manual derivation is tedious even for a motivated researcher. FullSimplify reveals that the sum of the squares of the magnitudes is not a function in z:

```
s=h3L*(h3L /. z→1/z)+h3H*(h3H /. z→1/z) // FullSimplify
```

which yields $2k0^2(1+k0^2)(1+k0^2)(1+k0^2)$.

The derived expression does not depend on z or b. Making it easy to draw schematics allows us immediately to become a researcher. We can do experiments with schematics, change them slightly, and re-run them. Very probably, we will end up doing calculations that have never been done before. Result that we get might become a discovery. For example, we can symbolically compute k0 for which s equals one

```
Solve[s==1, k0][[2]]
```

Out[57]=

$$\left\{k0 \to \frac{1}{\sqrt{2}\sqrt{1+k1^2}\sqrt{1+k2^2}\sqrt{1+k3^2}}\right\}$$

We know that h3L is zero for z=-1:

```
num3=Numerator[h3L//Together]/. z→1// Factor
```

which yields

$$-(1+b)^5\, k0\, (1 - k1 - k2 - k1\, k2 - k3 - k1\, k3 - k2\, k3 + k1\, k2\, k3)$$

Now, we can find k2 in terms of k1 and k3:

```
Solve[num3==0, k2] // Flatten // FullSimplify
```

Out[76]=

$$\left\{k2 \to \frac{-1 + k1 + k3 + k1\, k3}{-1 + k1\,(-1 + k3) - k3}\right\}$$

We derived the two expressions for k0 and k2 in terms of k1 and k3, directly from the schematic, using the two properties of the high-speed filter. We made no assumptions on the filter approximation function.

Design Equations

Third-order elliptic half-band IIR filter. The third-order elliptic half-band filter is the basic building block of the high-speed filter shown in Figure 16. It is important to implement the coefficient b with a small number of adders, for example `b=b0=1/2+1/24`, because it appears five times in the recursive path. Starting with `b`, we should find the passband edge frequency of the corresponding nonlinear-phase FIR filter [22]. The passband edge frequency of the FIR filter, `fpFIR`, can be computed from the known passband edge frequency of the IIR filter, `fp`, with the function `fp2fpFIR`:

```
fp2fpFIR[fp_] := Module[{b, delta2, L, r},
  b = fp2beta[fp];
  r = 2 (b (1 - b) + Sqrt[b (1 - b) (1 + 3 b)])/(1 + b)^2;
  L = Sqrt[(1 + 2 r)^3/((1 - r)^3 (1 + r))];
  delta2 = 1/Sqrt[1 + L];
  1/2 - ArcCos[delta2]/π]
```

$$r = 2\,\frac{b\,(1-b)+\sqrt{b\,(1-b)\,(1+3\,b)}}{(1+b)^2};\quad L=\sqrt{\frac{(1+2\,r)^3}{(1-r)^3\,(1+r)}};\quad \mathrm{delta2}=\frac{1}{\sqrt{1+L}};\quad \frac{1}{2}-\frac{\mathrm{ArcCos}[\mathrm{delta2}]}{\pi}$$

Note that `b` is also a function of `fp` [22]:

```
fp2beta[fp_] := Module[{a, b, r, x},
  a = 1/(Tan[π fp])^2; b = (4 (a^2 - 1)/a^4)^(1/3);
  r = 1/2 ((1 - 2/a^2)/Sqrt[1 + b + b^2] + Sqrt[2 + b + 2 Sqrt[1 + b + b^2]] - 1);
  x = Sqrt[1 + 2*r]/Abs[1 + r];
  (a + x^2 - Sqrt[(1 - x^2) (a^2 - x^2)])/(a + x^2 + Sqrt[(1 - x^2) (a^2 - x^2)])]
```

$$a=\frac{1}{(\mathrm{Tan}[\pi\,\mathrm{fp}])^2};\quad b=\sqrt[3]{\frac{4\,(a^2-1)}{a^4}};\quad r=\frac{1}{2}\left(\frac{1-\frac{2}{a^2}}{\sqrt{1+b+b^2}}+\sqrt{2+b+2\sqrt{1+b+b^2}}-1\right);\quad x=\frac{\sqrt{1+2*r}}{\mathrm{Abs}[1+r]};\quad \frac{\left(a+x^2-\sqrt{(1-x^2)\,(a\verb|^|2-x^2)}\right)}{\left(a+x^2+\sqrt{(1-x^2)\,(a\verb|^|2-x^2)}\right)}$$

We can compute fp for b=b0 and specified accuracy:

solF1=FindRoot[fp2beta[f]==b0,{f,0.1,0.2},AccuracyGoal→28, WorkingPrecision→32]

which gives {f→0.20464642998848691689350103435847}.

The script fpFIR=fp2fpFIR[f/.solF1] returns

fpFIR = 0.038167050863599031560177445517З. The fpFIR is the only parameter that we need in order to design the optimal half-band FIR filter of arbitrary order.

Optimal Third-Order Half-Band FIR Filter

Coefficients of the optimal nonlinear-phase third-order half-band FIR filter can be computed with the script

$$t = \mathrm{Cos}[2\,\pi\,\mathrm{fpFIR}];$$

$$a31 = \frac{3}{4}\,\frac{-9\,t - 9\,t^2 + 2\sqrt{3\,(1 + t + t^2)^3}}{(-2 + t + t^2)^2}$$

$$a33 = \frac{-9 + \sqrt{3\,(3 - 4\,a31)^3} + 18\,a31 - 6\,a31^2}{-36 + 54\,a31}$$

$$K32 = -\frac{1}{\sqrt{3}}\left(\left(-\sqrt{-\frac{a31}{a33}}\right) + \sqrt{3 - \frac{a31}{a33}}\right)$$

$$K31 = \sqrt{1 - a31/(3\,a33)} + K32$$

If set k1=K32, k2=K32, and

$$k0 = \frac{1}{\sqrt{2}\,\sqrt{1 + K31^2}\,\sqrt{1 + K32^2}}$$

The outputs `Y2L` and `Y2H` of the filter correspond to the lowpass and highpass high-speed filters based on the third-order IIR and the third-order FIR filters.

Optimal Fifth-Order Half-Band FIR Filter

Parameters of the optimal nonlinear-phase fifth-order half-band FIR filter can be computed in terms of r, which can be computed for known fpFIR with the script

$$\text{xp[r\_] := Module}\Big[\{P, q\},$$
$$q = \frac{3}{5} + \frac{2\,r}{5}\,\frac{(2+\sqrt{r})}{(1+2\sqrt{r})};\ P = \frac{r}{q};$$
$$\frac{\sqrt{q}}{3}\left(-2 + \sqrt[3]{5\left(6P - 1 - \sqrt{-5P^3 + (6P-1)^2}\right)} + \sqrt[3]{5\left(6P - 1 + \sqrt{-5P^3 + (6P-1)^2}\right)}\right)\Big]$$

```
solr = FindRoot[xp[r] == Cos[2 fpFIR π],
  {r, (Cos[fpFIR π])^2, 1},
   AccuracyGoal → 28, WorkingPrecision → 24]
```

which yields {r→0.957611623263286897864446}.
The coefficients are also functions in r:

$$R = \sqrt{r}\ \text{/. solr}$$
$$a = 1 + 2R - \sqrt{\left(\frac{1}{1+2R}\left(-1 - 2*R + 4R^2 + 4R^3 + 4R\sqrt{R\,(1+R)\,(1+3R+R^2)}\right)\right)}$$
$$K53 = \frac{a}{2\,(1+2R) - a}$$
$$K51 = (1+2R) - a$$

$$K52 = \frac{-1 + K51 + K53 + K51\,K53}{-1 - K51 - K53 + K51\,K53}$$
$$K50 = \frac{1}{\sqrt{2}\,\sqrt{1+K51^2}\,\sqrt{1+K52^2}\,\sqrt{1+K53^2}}$$

If we set k1=K51, k2=K52, k3=K53, and k0=K50, the outputs Y3L and Y3H correspond to the lowpass and highpass high-speed filters based on the third-order IIR and the fifth-order FIR filters. Here we use the new expressions for k0 and k2, in terms of k1 and k3, derived directly from the schematic, instead of the expressions reported in [22].

Magnitude and Phase Characteristics

DiscreteSystemFrequencyResponse computes and plots the magnitude and phase characteristics of a system:

{magnitude,phase}=DiscreteSystemFrequencyResponse[h2L/.parameterSubst3];

Specify numerical parameter values:

parameterSubst3={b→9/16, k0→0.346059527,k1→1.6990067, k2→-0.272445497}.

Here is the magnitude response:

```
magnitude // FullSimplify // ComplexExpand
```

$$\left(\sqrt{\left(\left(-68.7318+207.602\,\mathrm{Cos}[2 f \pi]+988.084\,\mathrm{Cos}[4 f \pi]+2272.37\,\mathrm{Cos}[6 f \pi]+3280.58\,\mathrm{Cos}[8 f \pi]+3463.29\,\mathrm{Cos}[10 f \pi]+2751.75\,\mathrm{Cos}[12 f \pi]+1714.55\,\mathrm{Cos}[14 f \pi]+761.993\,\mathrm{Cos}[16 f \pi]+252.277\,\mathrm{Cos}[18 f \pi]\right)^2+\left(207.602\,\mathrm{Sin}[2 f \pi]+988.084\,\mathrm{Sin}[4 f \pi]+2272.37\,\mathrm{Sin}[6 f \pi]+3280.58\,\mathrm{Sin}[8 f \pi]+3463.29\,\mathrm{Sin}[10 f \pi]+2751.75\,\mathrm{Sin}[12 f \pi]+1714.55\,\mathrm{Sin}[14 f \pi]+761.993\,\mathrm{Sin}[16 f \pi]+252.277\,\mathrm{Sin}[18 f \pi]\right)^2\right)}\right) / \left((9.+16.\,\mathrm{Cos}[4 f \pi])^2+256.\,\mathrm{Sin}[4 f \pi]^2\right)^{3/2}$$

This plots the magnitude characteristics, see Figure 17:

DiscreteSystemMagnitudeResponsePlot[h3L /.parameterSubstitution,

{0, 0.5},PlotRange→{-65,1}, PlotLabel→"Lowpass output",
AxesLabel→{"f","Gain (dB)"}];
parameterSubstitution={b→9/16,
k0→0.240006851761, k1→2.374283612,
k2→-0.540683329, k3→0.109326832}

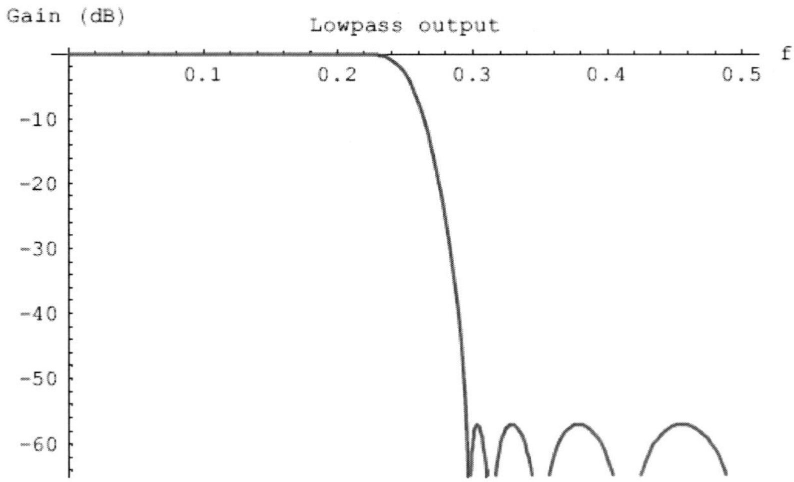

Figure 17. Magnitude characteristic of the high-speed filter.

DiscreteSystemImplementation creates a function that implements the high-speed filter:
DiscreteSystemImplementation[hsSystem,"hsf"];

The function name is hsf. SchematicSolver automatically generates the usage instruction:

{{Y9p8, Y9p0, Y20p8, Y20p0, Y31p8, Y31p0},{Y4p5, Y4p3, Y2p8, Y11p5, Y11p3, Y15p5, Y15p3, Y9p0, Y13p0, Y22p5, Y22p3, Y26p5, Y26p3, Y20p0, Y24p0}}=hsf[{Y0p8},{Y4p7, Y4p5, Y6p0, Y11p7, Y11p5, Y15p7, Y15p5, Y13p0, Y17p0, Y22p7, Y22p5, Y26p7, Y26p5, Y24p0, Y28p0}, {b, k0, k1, k2, k3}] is the template for calling the procedure.

The general template is {outputSamples,finalConditions} = procedureName[inputSamples, initialConditions, systemParameters].

See also: DiscreteSystemImplementationProcessing

The command ??hsf prints the function code:

```
hsf[dataSamples_List, initialConditions_List, systemParameters_List]:=
Module[{Y0p8, Y4p7, ... , Y28p5, Y31p8, Y31p0, Y26p3, b, k0, k1, k2,
k3},
{b, k0, k1, k2, k3} = systemParameters;
{Y0p8} = dataSamples;
{Y4p7, Y4p5, Y6p0, Y11p7, Y11p5, Y15p7, Y15p5, Y13p0, Y17p0,
Y22p7, Y22p5, Y26p7, Y26p5, Y24p0, Y28p0} = initialConditions;
Y6p3 = k1 Y6p0; Y2p8 = k0 Y0p8;
Y3p10 = Y2p8 + Y4p7;
Y3p6 = b Y3p10;
...
Y26p3 = Y24p8 - Y25p6;
{{Y9p8, Y9p0, Y20p8, Y20p0, Y31p8, Y31p0}, {Y4p5, Y4p3, Y2p8,
Y11p5, Y11p3, Y15p5, Y15p3, Y9p0, Y13p0, Y22p5, Y22p3, Y26p5,
Y26p3, Y20p0, Y24p0}}]
```

The above code is a software implementation, which is a sequence of statements that is executed by Mathematica. A similar code can be generated for executing on a general-purpose computer or on a dedicated hardware.

Symbolic Processing and Simulation

Let us process a unit impulse sequence of length 20:

```
numberInSamples = 20;
inputSequence=
UnitImpulseSequence[numberInSamples];
```

Assume zero initial conditions

```
eqns =
DiscreteSystemImplementationEquations[hsSystem];
initialConditions = 0*eqns[[2]]
```

```
Out[68]=
{0, 0, 0, 0, 0, 0, 0, 0, 0, 0, 0, 0, 0, 0, 0}
```

and identify the symbolic parameters:

systemParameters = eqns[[3]]

Out[69]=

```
{b, k0, k1, k2, k3}
```

Process the input sequence as follows:

{outputSeq, finalCond} = DiscreteSystemImplementationProcessing[inputSequence, initialConditions,systemParameters, hsf];

Output samples are expressions in terms of symbolic system parameters, e.g. the last sample at Y3L is
outputSeq[[numberInSamples,5]]// FullSimplify

Out[85]=

$$-b^5 (-1 + b^2)\, k0\, (b (7 - 9 b^2)\, k2 + k1 (-56 + b (-7\, k2\, k3 + b (280 + b (220 b (-2 + b^2) + 9\, k2\, k3)))))$$

Processing of a symbolic sequence results in a symbolic output sequence that is in terms of symbolic system parameters; this is a unique feature of SchematicSolver. For numerical parameter values, we obtain the simulation results as shown in Figure 18.

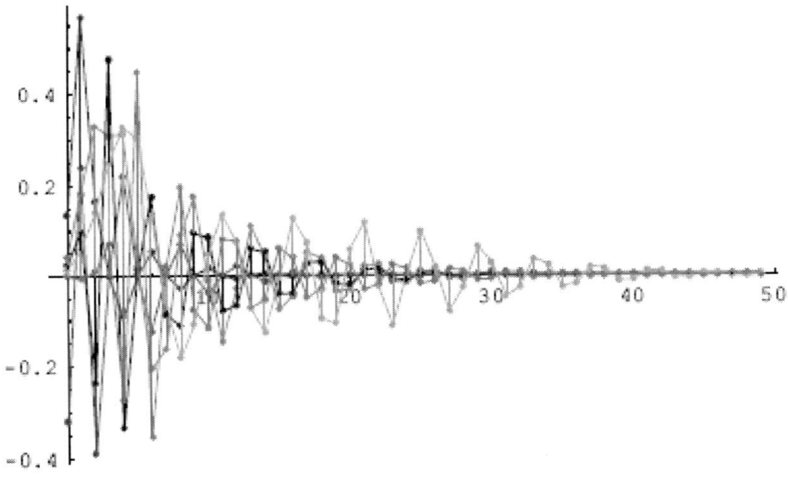

Figure 18. Impulse response of the high-speed filter.

Example 3 - Post-Processing with Advanced Symbolic Simulation

The development of modern electronic devices requires usage of very sophisticated and complex algorithms that combine expertise in the field of signal processing and software engineering. Classic computer tools that are based only on the numerical computations may be ineffective because the design space is almost unlimited. In this part of section, we present the possibility of computer algebra systems for symbolic processing, analysis and design of data obtained by numerical software tools. The symbolic processing using computer algebra systems is called post-processing.

The superiority of computer algebra systems in the development of signal processing algorithms can be demonstrated on digital filters [2], [11], [14], [23], [24], [25]. In the filter theory, the continuous time filter is defined as frequency selective electronic circuit that is used to amplify or attenuate sinusoidal signals [6]. Frequency range in which the signals are passed without significant distortion is called the passband. Frequency range in which the sinusoidal signals are rejected is called the stopband. The required minimum and maximum gain or attenuation, with appropriate passband and stopband edge frequencies is called the filter specification.

The digital filter is a discrete-time system that transforms a discreet-time signal composed of a sum of sinusoidal components into a new discrete-time signal in which the magnitude and phase of each individual sinusoidal component is changed. The amount of signal distortion depends on employed filter specification. Discrete-time signals are represented as a sequence of numbers. The improper usage of the number of samples or sampling frequencies may produce wrong decisions that is usually cause by spectrum leakage, folding or aliasing effects.

The computer algebra systems and symbolic processing can gain better insight into how some system works and its functionality and provide low-complexity implementation. Some computer algebra systems tools make system models as visualized algorithm using block diagram. The base model is a symbolic object that contains all the details of the drawing, simulation, and implementation of the system. The computer algebra systems can analyse continuous-time or digital system as a symbolic object with symbolic system parameters. Knowledge embedded in the object is used to the demanding task, to generate the implementation code or to identify the possible implementation problem.

Numerical Processing

Standard procedure for testing a new algorithm is to analyze a system, say a digital filter described by vectors num=[1 1 1] and den=[1 0.1 0.81]. Data in vector x is filtered using command y=filter(num, den, x) and the filtered data are in vector y. The spectrum of signal can be computed using discrete Fourier transform, i.e. using the command fft. Using the proper normalization, the spectrum of sinusoidal signal is shown in Figure 19. The spectrum computed using the fist 900 samples clearly shows that is the spectrum of the sinusoidal signal whose normalized frequency is 1/3. From the last 1000 samples it seems that there are a number of spectral components. Spectrums are quite different due to spectrum leakage.

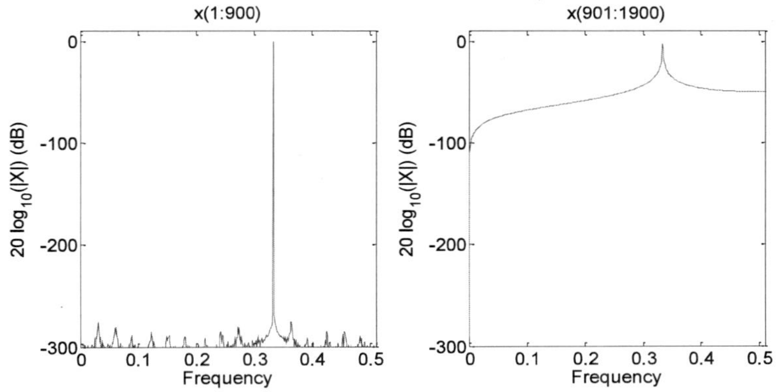

Figure 19. Spectrum of the sinusoidal sequence sin(2πfn), f=1/3, (a) the first 900 samples, (b) the last 1000 samples.

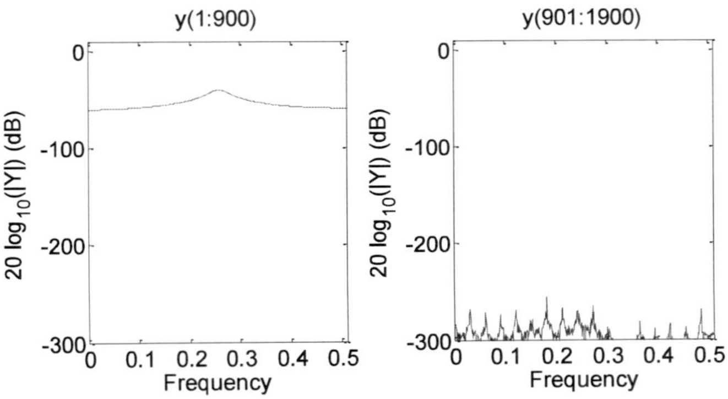

Figure 20. Spectrum of the filtered sequence y=filter(num,den,x), (a) the first 900 samples, (b) the last 1000 samples.

The spectrum of the filtered signal may be very confusing for inexperience users. It seems that the spectrum of the first 900 samples has many components that do not exist at the filter input. The spectrum computed from the last 1000 samples is quite different from the spectrum of the first 900 samples.

The first 900 filtered samples contain filter response due to zero initial states of delay elements; this is no response of the input signal. Also, it is not clear that the spectrum of the last 1000 samples is the result of quantization, see Figure 20.

Better insight in the functionality of a system can be obtained using computer algebra systems.

Built-in Collection of Schematics

In this example, SchematicSolver has a collection of functions that generate schematics of some systems that can be found in practice. The general format for calling the function is

{schematicSpec,inpCoords,outCoords}=SchematicAlbumFunction[params,{x0,y0},options]

SchematicAlbumFunction is the name of the function, e.g., DirectFormFIRFilterSchematic,

params is a list of parameters, say the values of filter coefficients.

{x0,y0} are the coordinates of the schematic. Typically, {x0,y0} are the coordinates of a system input.

schematicSpec is a schematic specification that represents the system; it is a list of element specifications.

inpCoords is a list of input coordinates.

outCoords is a list of output coordinates.

SchematicAlbumFunction generates a specification without input and output elements so that the output of any schematic can be connected to input of another schematic.

Some of the most important schematics are:

DirectFormFIRFilterSchematic creates schematic of Direct Form FIR filter.

HalfbandDirectFormFIRFilterSchematic creates schematic of Halfband Direct Form FIR filter.

HighSpeedIIR3FIRHalfbandFilter Schematic creates schematic of High-Speed 3rd-order IIR-FIR Halfband Filter.

HilbertTransformerDirectFormFIRSchematic creates schematic of the Hilbert Transformer with Direct Form FIR filter.

TransposedDirectForm2IIRFilter Schematic creates schematic of Transposed Direct Form 2 IIR filter.

After starting software Mathematica, the knowledge from SchematicSolver is activated using the command

Needs["SchematicSolver`"];

The digital filter described by vectors num=[1 1 1] and den=[1 0.1 0.81] is generated by calling built-in schematic. This is IIR filter. We create the schematic filterSchematic of Transposed Direct Form 2 IIR filter by adding input X and output Y to the built-in schematic, Figure 21.

```
{filterSchematic,inpCoords,outCoords}=
TransposedDirectForm2IIRFilterSchematic[{{b0, b1, b2}, {a1, a2}}];
mySchematic=Join[filterSchematic,
{{"Input", First[inpCoords], X}},
{{"Output", First[outCoords], Y}}];
```

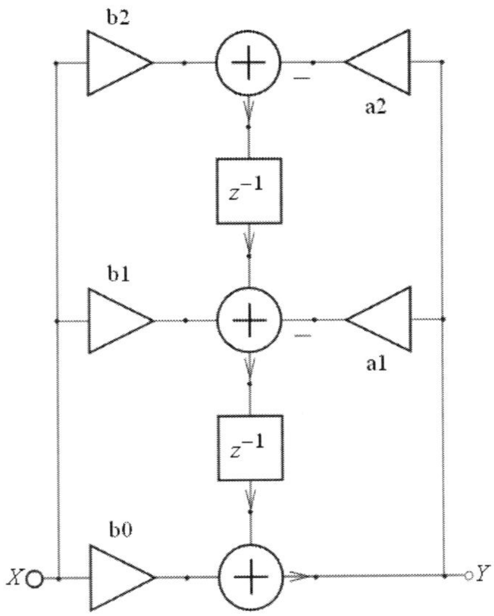

Figure 21. Automatically generated schematic of IIR filter.

The first difference in computer algebra systems with respect to the numerical processing is that all parameters are specified as symbols instead of numbers. The second difference is in automatically generated schematic in such a way that the number of elements depends on the number of parameters in the lists {{b0, b1, b2}, {a1, a2}}. Before filtering, an implementation function should be generated by clicking a button *Implement* on the corresponding pallet. Data in the input vector x is also created using predefined function for 1900 samples of the sinusoidal signal whose normalized frequency is 1/3: UnitSineSequence[1900, 1/3]. The filtered data are in vector y is computed using automatically generated implementation function from the schematic. For example, Figure 22 shows automatically generated implementation code and processing for 3 successive samples.

```
procedureName=implementationProcedure;
DiscreteSystemImplementation[mySchematic,
ToString[procedureName]];
inputSequence={{Sin[2 π n f ]}, {Sin[2 π (n+1)f]},{Sin[2 π (n+2)f]}};
initialConditions={q1, q2};
systemParameters={a1,a2,b0,b1,b2};
{outputSequence, finalConditions}=
DiscreteSystemImplementationProcessing [inputSequence,
initialConditions, systemParameters, procedureName];
```

Figure 22. Automatically generated implementation function.

The system parameters {a1,a2,b0,b1,b2}, the initial states of delay elements {q1,q2}, and the sample values {Sin[2π n f]} are presented using symbols. The symbols are set to some numerical values only when it is necessary, for example to draw the frequency response (Figure 23.)

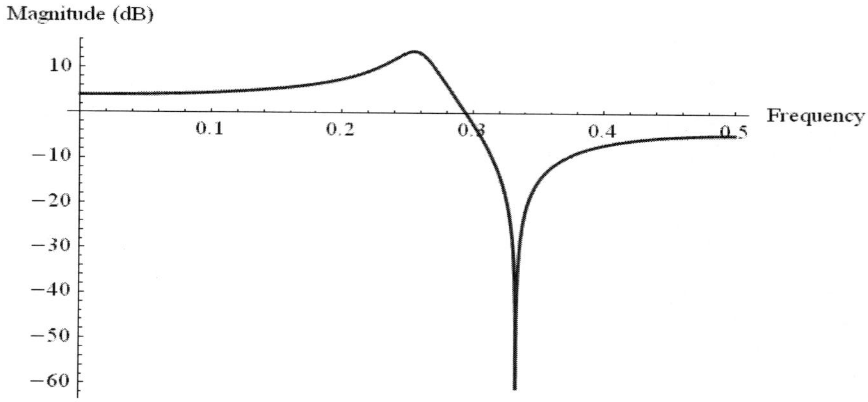

Figure 23. Frequency response.

```
myVal={b2→1,b1→1,b0→1,a1→1/10,a2→81/100};
DiscreteSystemFrequencyResponse[myTF/.myVal];
```

The spectrum of all signals can be computed using discrete Fourier transform, see Figure 24.

outputSequence[[Range[1,900]]] outputSequence[[Range[901,1900]]]

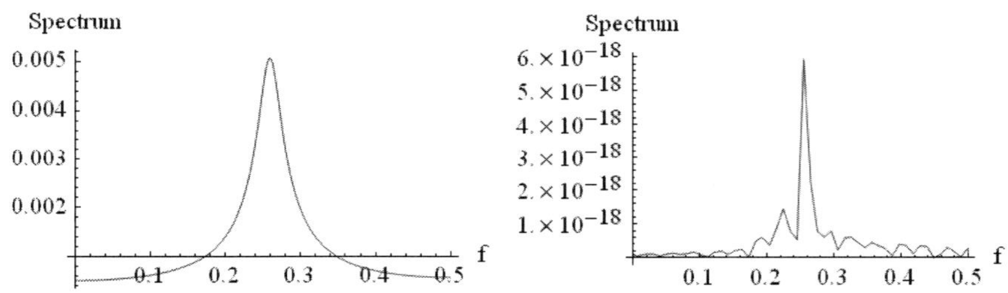

Figure 24. Spectrum of the filtered sequence in SchematicSolver, (a) the first 900 samples, (b) the last 1000 samples.

Notice that the results are generated using numerical processing. The spectrum of the filtered signal may be very confusing for inexperience users, again. It seems that the spectrum has many components that do not exist at the filter input.

The first examination is to derive the transfer function from the same schematic used for generating the implementation function

{myTF, Xinput, Youtput)=DiscreteSystemTransferFunction[mySchematic]

$$\left\{\left\{\left\{-\frac{-b2 - b1\, z - b0\, z^2}{a2 + a1\, z + z^2}\right\}\right\}, \{Y[\{0, 0\}]\}, \{Y[\{6, 0\}]\}\right\}$$

myTF // DiscreteSystemDisplayForm

$$\frac{b0 + b1\, z^{-1} + b2\, z^{-2}}{1 + a1\, z^{-1} + a2\, z^{-2}}$$

From the derived transfer function, and for b2=b1=b0=1, a1=0.1, a2=0.81, it follows that the sinusoidal signal whose frequency is 1/3 should be rejected by the filter. The spectrum obtained by numerical processing does not approve this expectation.

Figure 22 shows automatically generated implementation code and processing for 3 successive symbolic samples. The corresponding filtered samples are yn, yn1, yn2:

yn = outputSequence[[1]][[1]]
yn1 = outputSequence[[2]][[1]]
yn2 = outputSequence[[3]][[1]]
For example, yn1 becomes

```
q2 + Cos[2 f n π] Sin[2 f π] + 1/10 (-q1 - Sin[2 f n π]) +
  Sin[2 f n π] + Cos[2 f π] Sin[2 f n π]
```

To find the response that does not depend on initial states q and q2 of two delay elements, from two equations we find the values of q and q2. Also, we replace symbolic value of nth sample as a function `y[n]`:

```
sol1=Solve[yn==y[n],q1]//Flatten
sol2=Solve[yn1==y[n+1],q2]//Flatten
eq1=yn2→y[n+2]/.sol2/.sol1
```

The equation `eq1` becomes

```
Cos[2 f n π] Sin[2 f π] + Cos[2 f n π] Sin[4 f π] + Sin[2 f n π] +
  Cos[2 f π] Sin[2 f n π] + Cos[4 f π] Sin[2 f n π] -
  81 y[n]/100 + 1/10 (-Cos[2 f n π] Sin[2 f π] -
    Sin[2 f n π] - Cos[2 f π] Sin[2 f n π] + y[n]/10 +
    1/10 (10 Cos[2 f n π] Sin[2 f π] + 10 Sin[2 f n π] + 10 Cos[
        2 f π] Sin[2 f n π] - y[n] - 10 y[1 + n])) == y[2 + n]
```

This is a recurrence equation. It can be solved using a command for solving recurrence equation `RSolve`.

```
sol=RSolve[eq1,y[n],n]/.C[1]→0/.C[2]→0;
```

The solution is very complex and it is not shown. The derived symbolic value that is a function of the signal frequency `f` and the index `n` can be shown for specified frequency, say f=1/3

```
In[30]:= y[n] /. sol /. f→1/3 // Simplify
Out[30]=
{0}
```

The result is in Output[30] and it is a 0. This proves that the sinusoidal signal whose frequency is 1/3 cannot pass the filter. Therefore, the spectrum in Figure 24 of the filtered signal contains components caused by the initial states of the delay elements and quantisation effects. Notice that the zero value of the filtered signal is derived using implementation function and that it is not a result of analysis of transfer function. The filtered signal can be also derived for some other frequency, say

```
f=1/3 y[n]/. sol/. f→3/10//Simplify
```

The final result shows that the filtered signal is the attenuated sinusoidal signal with phase shift:

```
{-0.2833 Cos[1.885 n] - 0.72013 Sin[1.885 n]}
```

Post-Processing

SchematicSolver generates software implementation of a system directly from the system schematic. SchematicSolver may process data samples symbolically, keeping the system parameters as symbols. The unknown system parameter can be identified at the end of processing. For example, if the parameter b1 is not set to any numerical value, the recurrence equation is shown in Figure 25. Specifying the zero value of the filtered signal, y[n]==0, then b1 is computed to be 1.

```
sol=RSolve[eq, y[n], n] /. C[1]→0 /. C[2]→0;
x=Solve[(y[n] /. sol)==0, b1]

{{b1 → 1}}
```

$$-\frac{81\,y[n]}{100}-\frac{1}{2}\sqrt{3}\,\text{Cos}\left[\frac{2\,n\,\pi}{3}\right]+$$
$$\text{b1}\left(\frac{1}{2}\sqrt{3}\,\text{Cos}\left[\frac{2\,n\,\pi}{3}\right]-\frac{1}{2}\,\text{Sin}\left[\frac{2\,n\,\pi}{3}\right]\right)+\frac{1}{2}\,\text{Sin}\left[\frac{2\,n\,\pi}{3}\right]+$$
$$\frac{1}{10}\left(\frac{y[n]}{10}-\frac{1}{2}\sqrt{3}\,\text{Cos}\left[\frac{2\,n\,\pi}{3}\right]+\frac{1}{2}\,\text{Sin}\left[\frac{2\,n\,\pi}{3}\right]-\text{b1}\,\text{Sin}\left[\frac{2\,n\,\pi}{3}\right]+\right.$$
$$\frac{1}{10}\left(-y[n]-10\,y[n+1]+5\sqrt{3}\,\text{Cos}\left[\frac{2\,n\,\pi}{3}\right]-5\,\text{Sin}\left[\frac{2\,n\,\pi}{3}\right]+\right.$$
$$\left.\left.10\,\text{b1}\,\text{Sin}\left[\frac{2\,n\,\pi}{3}\right]\right)\right)==y[n+2]$$

Figure 25. Recurrence equation of the filtered signal for unknown parameter b1 and the known frequency f=1/3.

Benefits of symbolic processing and the role of computer algebra systems can be highlighted from the viewpoint of both academia (derivation of time and frequency response, proving system properties) and industry (verification of realizations). For example, let us start with known IIR filter structure shown in Figure 21 and find the parameters of the system that has two identical real poles and minimal overshoot of the step response.

We create the schematic filterSchematic of Transposed Direct Form 2 IIR filter by adding input X and output Y to the built-in schematic, Figure 21.

The additional requirement is to have the step response 1 for large index n, and symmetric coefficients in the numerator of the transfer function. This means that b2 is the same as b0, and b0 can be expressed as a function of a1, a2, and b1. The two parameters a1 and a2 are function of real pole a0, that is

```
In[6]:= myTFn =
          myTF /. b2 → b0 /. b0 → 1/2 (1 + a1 + a2 - b1) /. a1 → 2 a0 /.
            a2 → a0^2 // Simplify // Flatten
```

The implementation function is generated by clicking a button *Implement* on the corresponding pallet. Data in the input signal x is created using predefined function of the step signal UnitStepSequence[10]. The filtered signal outputSequence is computed using automatically generated implementation function from the schematic. The unknown values are b1 and a0.

Suppose that we expect the second sample of the filtered signal to be 1, and find a0

In[20]:= y2=(outputSequence[[2]])[[1]] // FullSimplify

Out[20]=

$$\frac{1}{2}\,(1+b1+a0\,(-a0\,(3+2\,a0)+2\,b1))$$

In[21]:= sol1=(Solve[y2==1, a0] // Flatten) [[1]] // Simplify

Out[21]=

$$a0\to-\frac{1}{2}+\frac{\left(-162+2\sqrt{6561+(-9-12\,b1)^3}\right)^{1/3}}{6\,2^{1/3}}+\frac{3+4\,b1}{2^{2/3}\left(-162+2\sqrt{6561+(-9-12\,b1)^3}\right)^{1/3}}$$

The minimal overshoot can be found solving the equation

```
solb1all = Solve[6561 + (-9 - 12 b1)^3 == 0, b1]
% // N
```

$$\left\{\left\{b1\to-\frac{3}{4}+\frac{3\,3^{2/3}}{4}\right\},\left\{b1\to-\frac{3}{4}-\frac{3}{8}\,3^{2/3}\left(1-i\sqrt{3}\right)\right\},\left\{b1\to-\frac{3}{4}-\frac{3}{8}\,3^{2/3}\left(1+i\sqrt{3}\right)\right\}\right\}$$

$$\{\{b1\to 0.810063\},\{b1\to -1.53003+1.35105\,i\},\{b1\to -1.53003-1.35105\,i\}\}$$

The parameter should be real, that is

$$\left\{b1\to-\frac{3}{4}+\frac{3\,3^{2/3}}{4}\right\}$$

This is the value of b1 for the minimal overshoot. Finally, the plot is shown in Figure 26 using the corresponding plot command for computed values a1 and b0

```
SequencePlot[outputSequence/.sol1/.solb1];
```

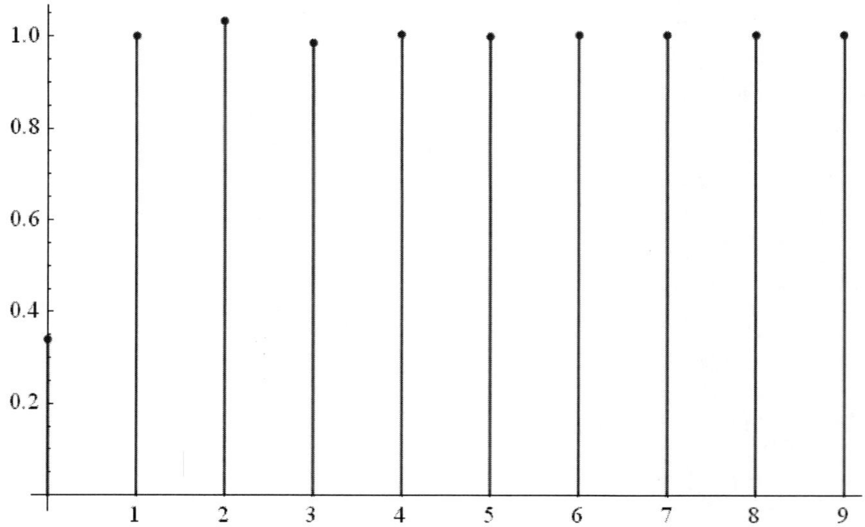

Figure 26. Step response with minimal overshoot.

Example 4 - Simulation of OFDM Transmitters and Post Processing

In this example we present an original step-by-step procedure for simulations of an OFDM (Orthogonal Frequency Division Multiplexing) transmitter using computer algebra systems and symbolic signal processing. The expert-based knowledge in the symbolic object was used to simulate an example system OFDM DVB (Digital Video Broadcasting) and to generate the implementation code of some critical parts of the system.

OFDM (Orthogonal Frequency Division Multiplexing) is a digital multicarrier transmission technique that distributes the digitally encoded symbols over several subcarrier frequencies in order to reduce the symbol clock rate and to achieve robustness against long echoes in a multipath radio channel [26], [27], [28]. The spectra of the individual subcarriers overlap and the information can be completely recovered without any interference from other subcarriers. From a mathematical point of view, this is a consequence of the orthogonality of the base functions of the Fourier series. This technique is very popular for professional applications and consumer electronics in wireless communication systems [26].

Many research centers in the world have specialized teams working in the optimization of OFDM for countless applications, one such group is the Georgia Institute of Technology – Smart Antenna Research Laboratory (SARL), a part of the Georgia Center for Advanced Telecommunications Technology (GCATT). The paper [29] provides Matlab [6] code that simulates the basic processing involved in the generation of an OFDM signal in a physical channel with detailed description of each of the simulation steps. Author expected that his paper [29] provided a better explanation of the fundamentals of OFDM. It is interesting that some signals of Matlab simulation and the direct simulation were similar but not identical for the same input data. Furthermore, [29] concludes that different results are expected for the same input data because they had used different FFT–IFFT blocks.

Unexpected results of numerical-only simulation force researcher to guess the phenomenon being investigated, instead to find the accurate solution. The use of computer algebra systems and symbolic processing can help to gain insight into how a complex system works, which is preferred to experimenting with numerical simulations. The term post-processing is used in the signal processing for usage of additional software that generates understandable results, which cannot be derived by former software. The purpose of this paper is to provide computer algebra systems code to simulate the basic processing involved in the generation of an OFDM signal in a physical channel and to find out the reasons for unexpected differences in signals obtained using numerical-only simulations. The computations are performed with Mathematica [5] and the application package SchematicSolver [23]. The simulation model is the same as presented in [29]. More details on how computer algebra systems can be used in control theory, telecommunications, numerical nonlinear optimization, and digital filter design can be found in [2], [14], [16], [25].

OFDM Transmitter

Simulation Model

The simulation model is based on the OFDM signals of the Digital Video Broadcasting (DVB) standard for the European terrestrial digital television (DTV) service [29]. The step-by-step procedure follows the Matlab code described in [29].

OFDM Symbol Generation

A detailed description of OFDM can be found in [28] where we can find the expression for the OFDM symbols. The first step is to load the package SchematicSolver in computer algebra systems Mathematica with the command in the working notebook

```
Needs["SchematicSolver`"]
```

The next step is to draw a block diagram of the OFDM symbol generation using point-and-click interface for drawing systems. A simplified model is shown in Figure 27.

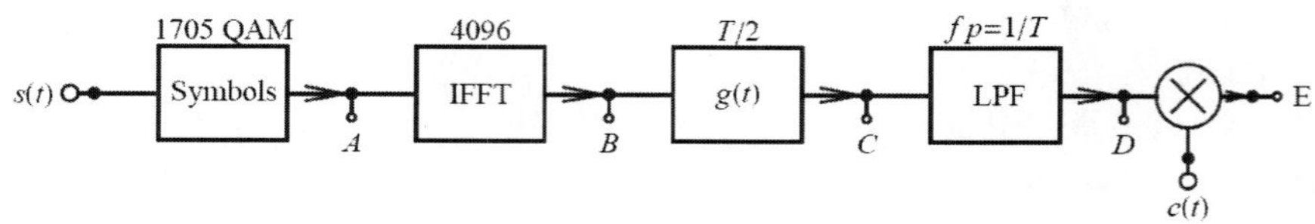

Figure 27. Block diagram of OFDM symbol generation.

The carrier frequency is close to 90 MHz, which is in the range of a VHF channel five, a common TV channel in any city. The appropriate simulation period T is defined as the elementary period for a baseband signal, and it relates to a time-period, 1/Rs. For simplicity, we use Rs=40/T.

DVB-T Parameters

Numerical values for the OFDM parameters for the 2k mode are specified as in [29]. Notice that all values are integer. The computer algebra system works with integer numbers as symbols and the final result is exact:

```
Tu = 224*10^-6;(* useful OFDM symbol period *)
T = Tu/2048; (* baseband elementary period *)
G = 0; (* choice of 1/4,1/8,1/16,and 1/32 *)
delta = G*Tu;(* guard band duration *)
Ts = delta + Tu;(* total OFDM symbol period *)
Kmax = 1705; (* number of subcarriers *)
Kmin = 0;
FS = 4096; (* IFFT/FFT length *)
q=10;(* carrier period to elementary period ratio *)
fc = q*1/T;(* carrier frequency *)
```

```
Rs = 4*fc;(* simulation period *)
M = Kmax + 1;
```

The OFDM spectrum is centered on carrier frequency [29]. One simple way to achieve the centering is to use a 2N-IFFT as in [28] and T/2 as the elementary period (N=2048). FFT is Fast Fourier Transform and IFFT is inverse FFT. Therefore, 4096-IFFT was used in [29].

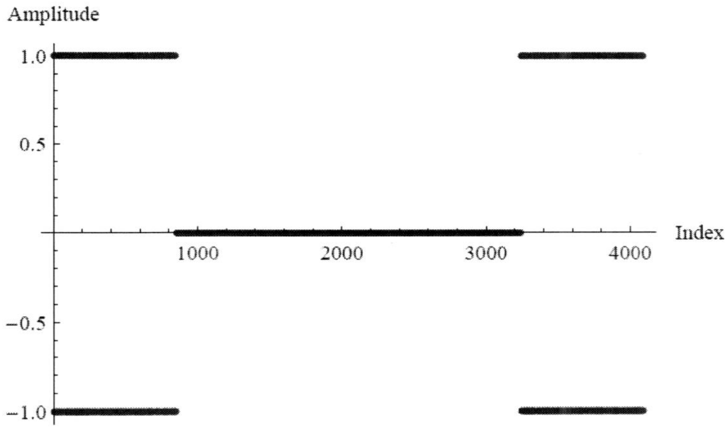

Figure 28. Information signal at A.

Information Signal

Usually, a random generator is used for simulating information signal. In this paper, we are using the same information signal generated in [29] using Matlab code, in order to compare signals obtained by Matlab and Mathematica, as two sequences an1 and an2. Alternatively, we can use the command Random to generate information signal in Mathematica.

The input signal a consists of 1705 symbols (+1 or -1) and it is shown in Figure 28. As suggested in [28], 2 391 zeros are added to the information signal at (A) to achieve over-sampling. Signal at B is computed using IFFT. SchematicSolver provide appropriate commands, so the code is very simple:

```
a = -1 + 2 an1 + I (-1 + 2 an2);
info = ListToSequence[Table[0, {i, 1, FS}]];
info[[1 ;; M/2]] = a[[1 ;; M/2]];
info[[-M/2 ;; -1]] = a[[M/2 + 1 ;; M]];
SequencePlot[Re[info][[1;;FS]],StemPlot→False,Joined→False,
AxesLabel→{"Index","Amplitude"}];
```

carriers=FS*InverseFourier[info,FourierParameters->{1,-1}];

The signal carriers use T/2 as its time period, as it is shown in Figure 29 for Carriers Inphase at B.

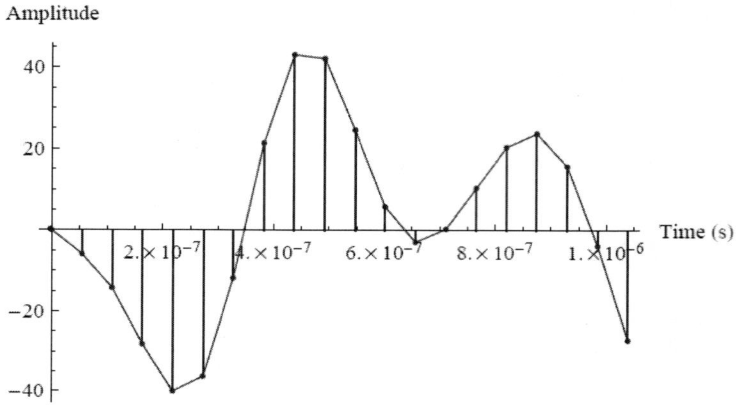

Figure 29. Carriers inphase at B.

Carriers FFT at B

Short-term FFT (SFFT) shows a large variation of the spectral components, see Figure 30:

spectrum = (1/FS) SequenceDiscreteFourierTransform[carriers];

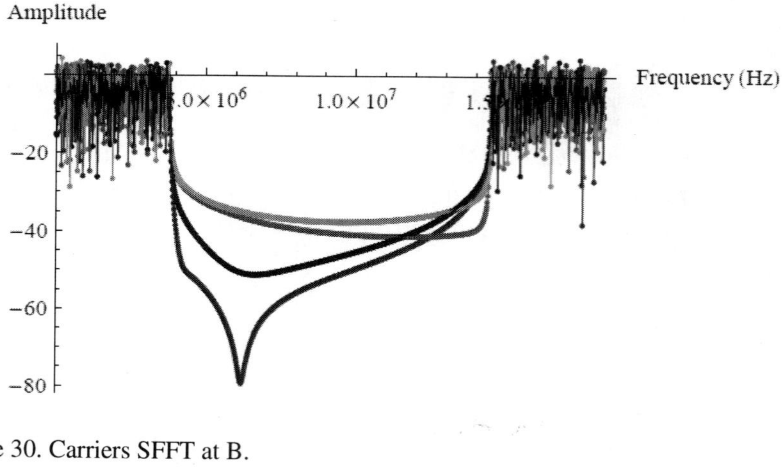

Figure 30. Carriers SFFT at B.

4096-FFT of the signal at B is shown in Figure 31. As it can be expected, it is similar to the magnitude of information signal.

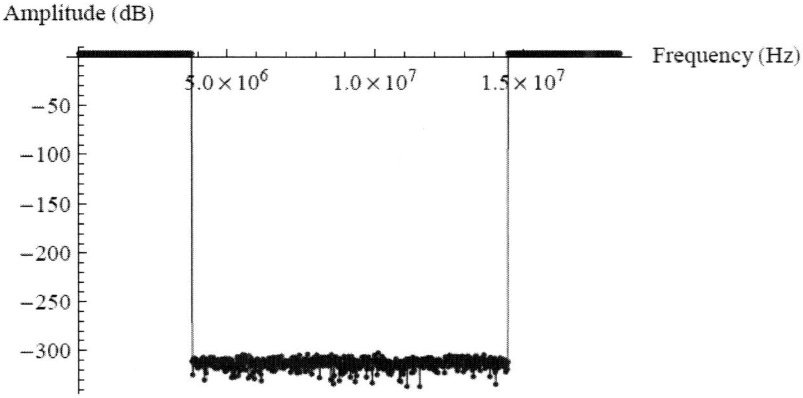

Figure 31. Carriers 4096-FFT at B.

Generation of Continuous Signal

The information signal is the discrete-time signal and the carriers at B should be also the discrete-time signal.

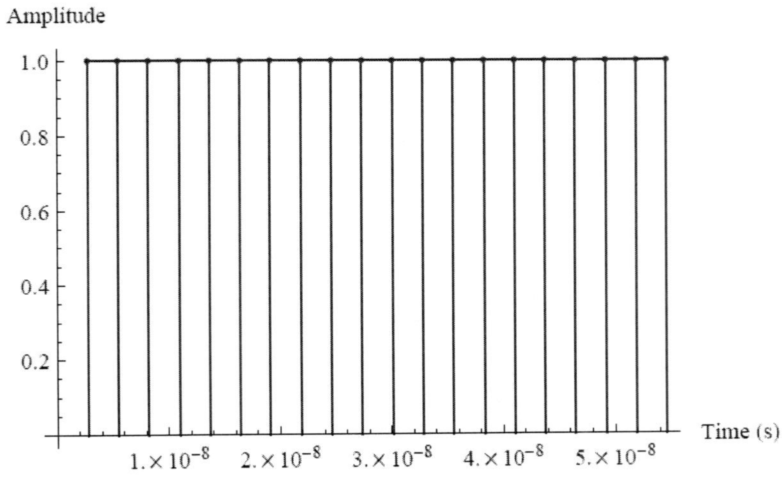

Figure 32. Pulse shape `g(t)`.

In order to simulate continuous-time signal, a pulse shape g(t) signal is generated, as it is shown in Figure 32:

```
g = ListToSequence[Table[1,{i,1/Rs,T/2,1/Rs}]];
```

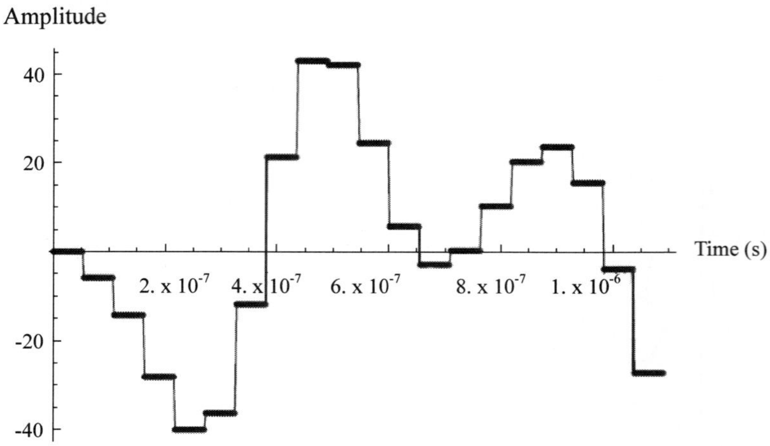

Figure 33. Inphase part of the signal u at C.

The signal u(t) at the output of pulse shaping block, at C, is up-sampled signal by a factor of 20, and filtered using finite impulse-response filter (FIR), as shown in Figure 33 for the real part of the signal. The SchematicSolver code follows:

```
u=ListToSequence[ListConvolve[SequenceToList[g],
SequenceToList[UpsampleSequence[carriers,20]],1]];
```

The spectrum of the signal u(t) shows very large number of images of the base-band signal, as it is shown in Figure 34. The next step is to reject images using a lowpass filter. Up to this step, all simulated signals are the same in Matlab and Mathematica.

In [29], the proposed reconstruction or "D/A filter" response is the 13th-order Butterworth filter with the cut-off frequency of approximately 1/T. Butterworth filters are characterized by a magnitude response that is maximally flat in the passband and monotonic overall. The magnitude response shown in [29] is noise-like with larger oscillations in the passband. A simulation in Matlab, using the function freqz, shows a large passband and stopband variations, as shown in Figure 35.

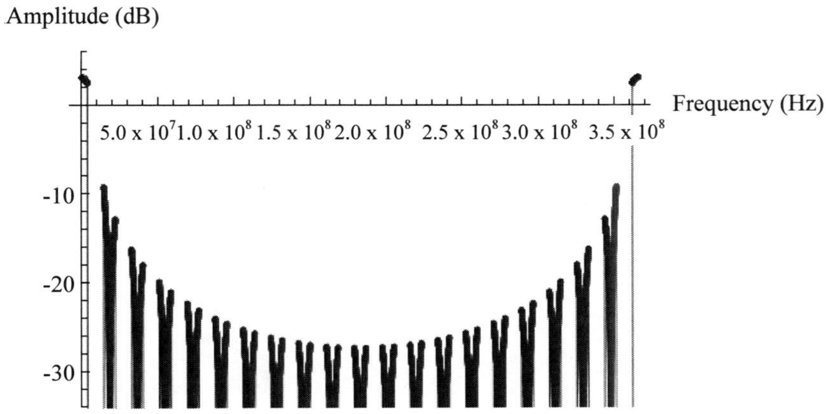

Figure 34. Frequency response of the signal u at C.

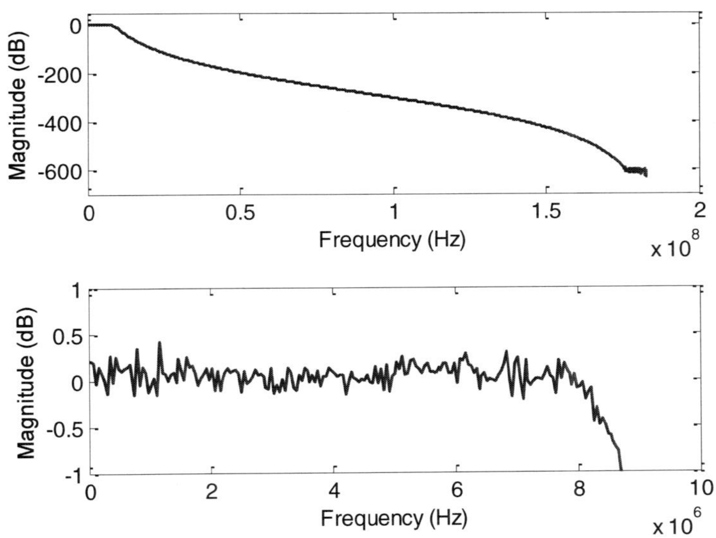

Figure 35. Frequency response of the13th-order Butterworth filter (Matlab).

Nevertheless, we can draw the filter using application package SchematicSolver, construct the digital filter and perform filtering. The filtered signals are shown in Figure 36. Signals are similar to the corresponding obtained in Matlab using function filter.

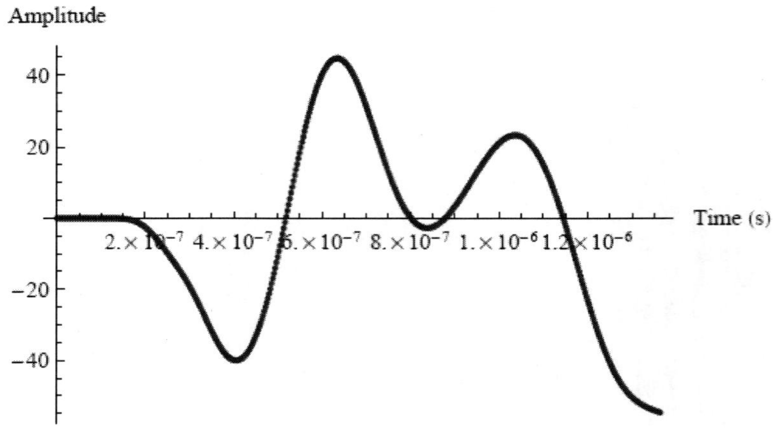

Figure 36. Inphase filtered signal at D.

The spectrum of the filtered signal shows that all spectral images are eliminated, as it is shown in Figure 37.

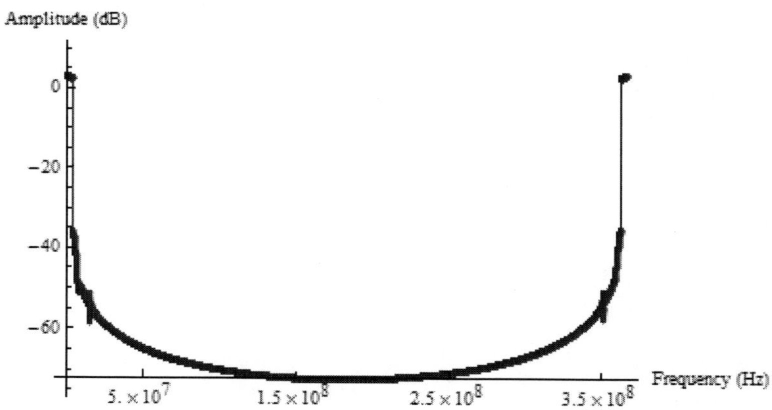

Figure 37. Frequency response of the filtered signal at D.

The next step is to perform the quadrature multiplex double-sideband amplitude modulation of filtered signal. In this modulation, the inphase signal and the quadrature signal are modulated using the code in Mathematica:

```
nn = Length[uoft];
s=Re[ListToSequence[SequenceToList[uoft[[1;;nn]]]*
 E^(I 2\[Pi] fc Table[i, {i, 0, nn-1}])]];
```

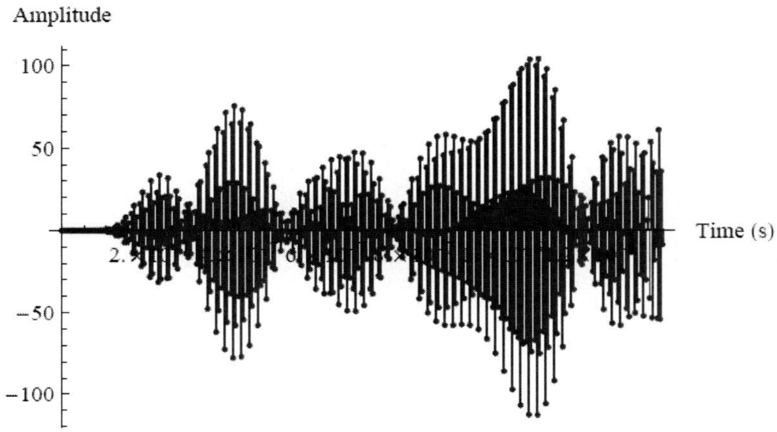

Figure 38. Transmitted signal at E.

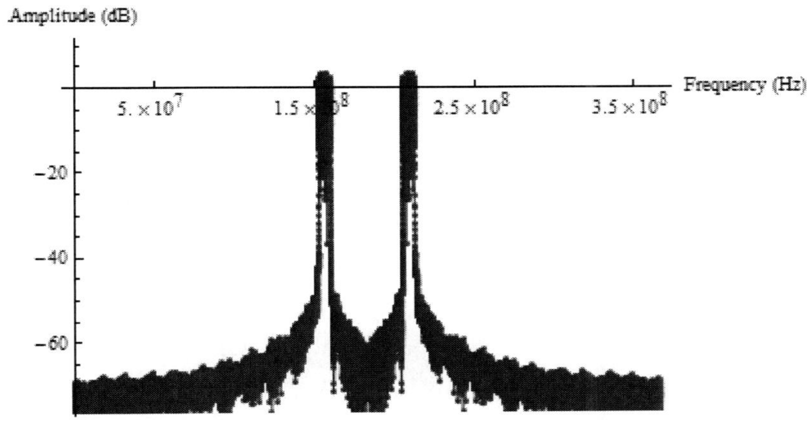

Figure 39. Frequency response of the transmitted signal at E.

Time response of signal `s(t)` at E is shown in Figure 38, while the frequency response is shown in Figure 39. SchematicSolver describes a system as a symbolic object in the form of a list of elements. Two different representations exist: the graphical representation of the block-diagram (as shown in Figure 40) and the textual description of the block-diagram.

Symbolic processing can be carried out as follows: automatically generate a code that implements the system, compute the input sequence whose elements can be symbols or numbers, and process the input sequence with the code that is the same for simulation and implementation.

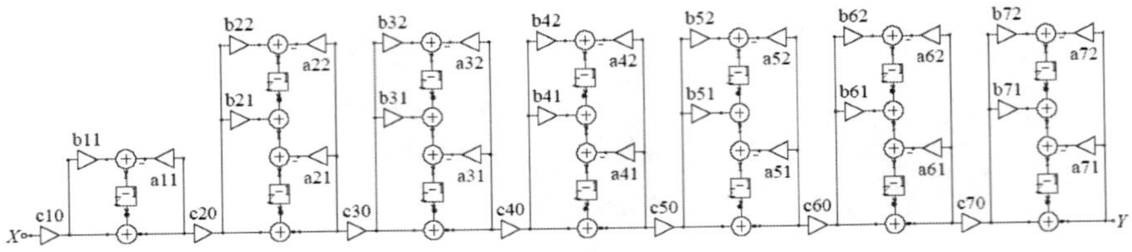

Figure 40. Schematic of the 13th-order Butterworth filter.

A variable precision can be used for computing the frequency response or filtering the signal. A direct form implementation of the filter has the same frequency response as in [29] for double precision. The frequency response of the filter shown in Figure 40 has no passband variations as in [16]. Therefore, a cascade connection of second-order sections is more appropriate for implementation than the direct form used in [16].

OFDM Receiver

As we mentioned before, the design of an OFDM receiver is open; i.e., there are only transmission standards. With an open receiver design, most of the research and innovations are done in the receiver. For example, the frequency sensitivity drawback is mainly a transmission channel prediction issue, something that is done at the receiver; therefore, we shall only present a basic receiver structure in this report.

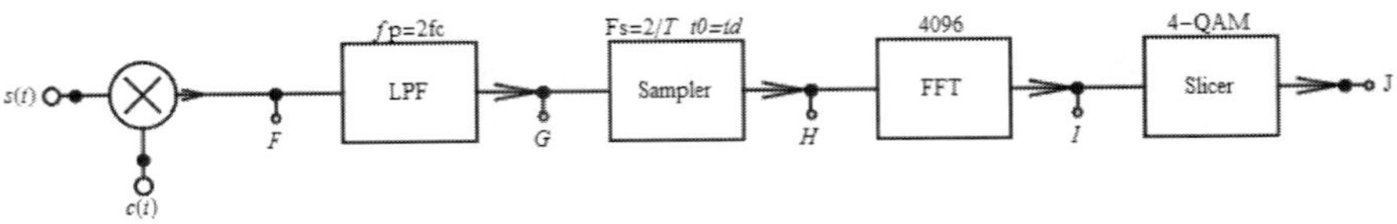

Figure 41. Block diagram of OFDM symbol receiver.

OFDM is very sensitive to timing and frequency offsets. Even in this ideal simulation environment, we have to consider the delay produced by the filtering operation. For our simulation, the delay produced by the reconstruction and demodulation filters is about td=64/Rs. This delay is enough to impede the reception, and it is the cause of the slight differences we can see between the transmitted and received signals. With the delay taken care of, the rest of the reception process is straightforward. As in the transmission case, we specified the names of the simulation variables and the output processes in the reception description. We will demonstrate step by step symbolic analyses as shown on Figure 41. DVB-T parameters are the same as OFDM transmitter.

UOFT Inphase

```
{uoftR,uoftFinalR}=DiscreteSystemImplementationPr
ocessing[Re[u],                   initialConditions,
systemParameters, procedureName];
SequencePlot[uoftR[[1;;500]],
SequenceSamplingFrequency→Rs,          Joined→True,
StemPlot→False,                       AxesLabel→{"Time
(s)","Amplitude"}]
```

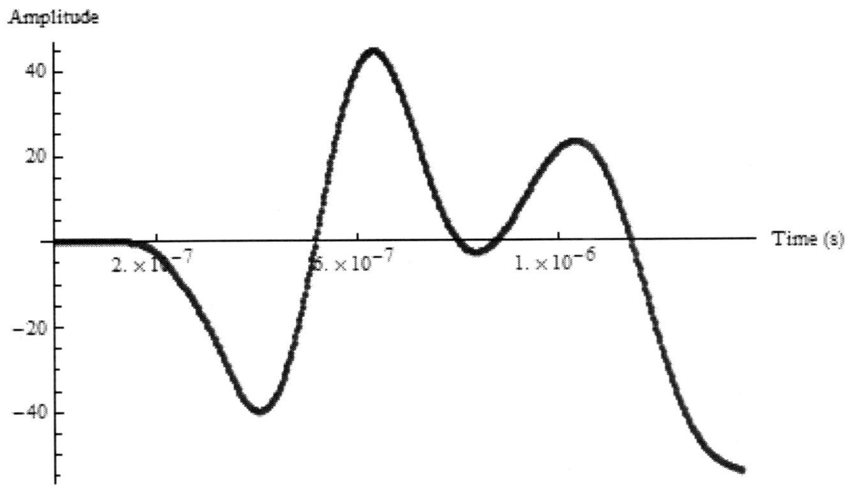

Figure 42. Time response of received signal in phase.

UOFT Quadrature

```
{uoftI,uoftFinalI}=DiscreteSystemImplementationPr
ocessing[Im[u],                   initialConditions,
systemParameters, procedureName];
SequencePlot[uoftI[[1;;500]],
SequenceSamplingFrequency→Rs,          Joined→True,
```

```
StemPlot→False,AxesLabel→{"Time
(s)","Amplitude"}]
```

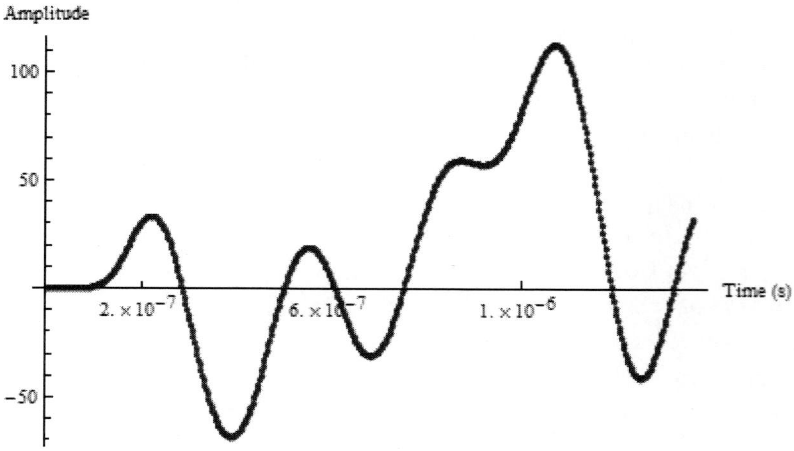

Figure 43. Time response of received signal in quadrature.

```
uoft=uoftR+I uoftI;
```

```
nn=Length[uoft];
s=Re[ListToSequence[SequenceToList[uoft[[1;;nn]]]
e^(I 2. π fc Table[i,{i,0,nn-1}])]];
```

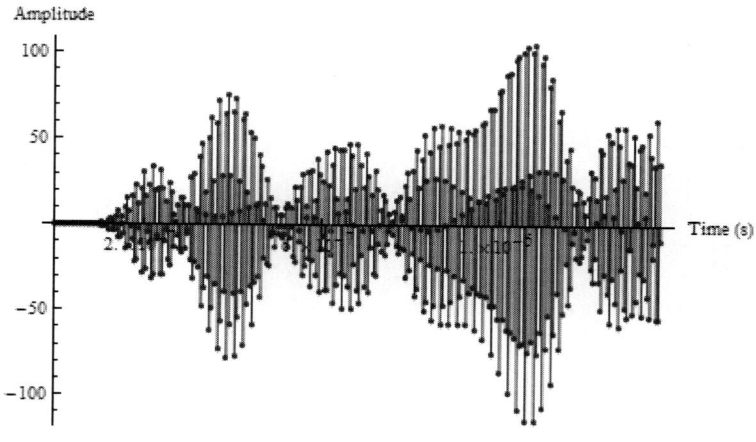

Figure 44. Original time response of received signal.

For our simulation, the delay produced by the reconstruction and demodulation filters is about td=64/Rs. This delay is enough to impede the reception, and it is the cause of the slight differences we can see between the transmitted and received signals [29].

```
delay=64;
uoftnew=PadLeft[uoft[[;;]]//
SequenceToList,Length[s]+delay]                     //
ListToSequence;
mm=Length[uoftnew];
s=Re[ListToSequence[SequenceToList[uoftnew[[1;;mm
]]] e^(I 2. π fc Table[i,{i,0,mm-1}])]];
SequencePlot[Re[s[[1;;500]]],
SequenceSamplingFrequency→Rs,AxesLabel→{"Time
(s)","Amplitude"}]
```

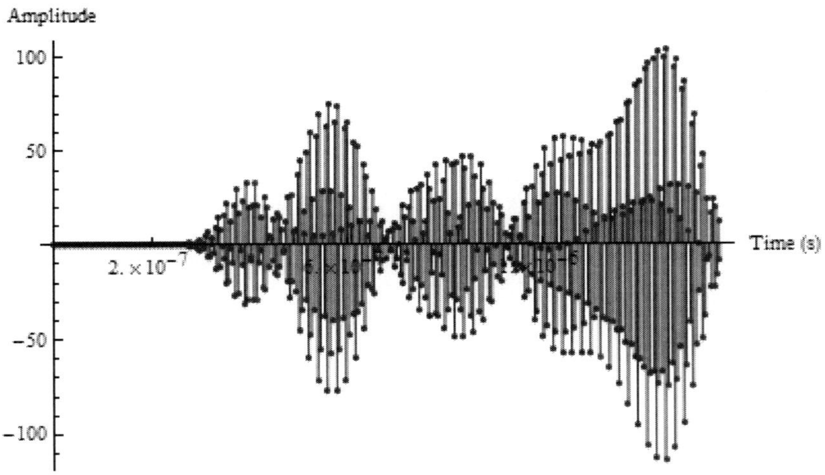

Figure 45. Time response of of received signal with time delay.

Practically, the process of receiving and processing of signal begins in point (F) – downconversion.

```
rTilde=ListToSequence[e^(     -I     2.     π     fc
Table[i,{i,0,nn-1}])*SequenceToList[s[[1;;nn]]]];
rTildeRe=Re[rTilde];
SequencePlot[rTildeRe[[1;;500]],
SequenceSamplingFrequency→Rs,          Joined→True,
```

```
StemPlot→False,AxesLabel→{"Time
(s)","Amplitude"}]
```

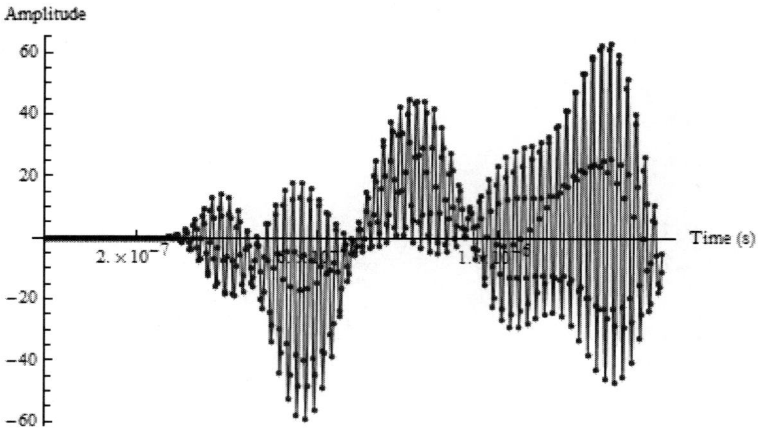

Figure 46. Time response of of real received signal with time delay.

```
rTildeIm=Im[rTilde];
SequencePlot[rTildeIm[[1;;500]],
SequenceSamplingFrequency→Rs,          Joined→True,
StemPlot→False,AxesLabel→{"Time
(s)","Amplitude"}]
```

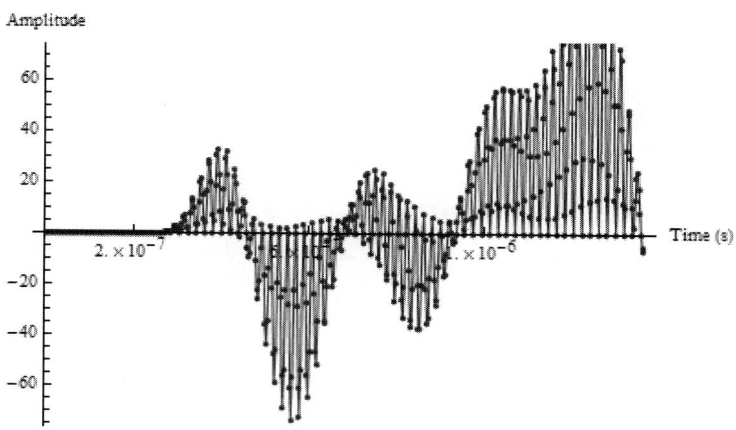

Figure 47. Time response of of imaginary received signal with time delay.

Extraction of carriers is done using FFT.

```
norm=Length[rTilde];
spectrumrTilde=1/norm
SequenceDiscreteFourierTransform[rTilde];
SequencePlot[20 Log[10,Abs[spectrumrTilde]],
Joined→True, StemPlot→False,
SequenceSamplingFrequency→Tu,AxesLabel→{"Frequen
cy (Hz)","Amplitude (dB)"}, PlotRange→{{-
10,1.5*10^8},{-120,6}}]
```

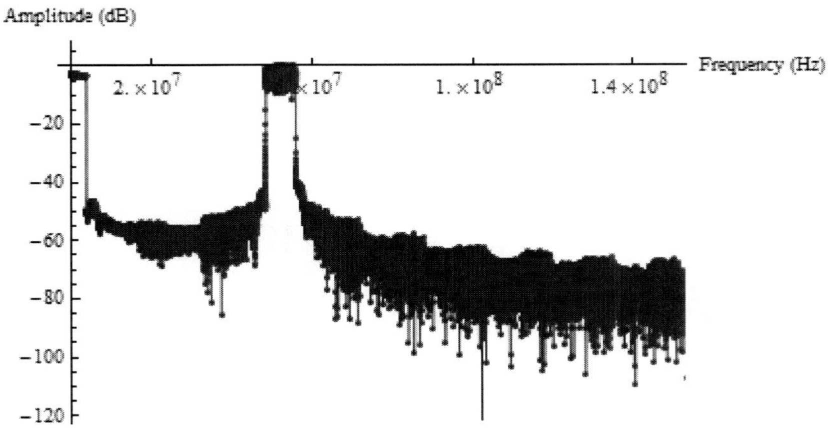

Figure 48. Frequency response of signal at (F).

```
rInfoRe,
rInfoFinalR}=DiscreteSystemImplementationProcessi
ng[rTildeRe, initialConditions, systemParameters,
procedureName];
SequencePlot[rInfoRe[[1;;500]],
SequenceSamplingFrequency→Rs, Joined→True,
StemPlot→False,AxesLabel→{"Time
(s)","Amplitude"}]
{rInfoIm,
rInfoFinalI}=DiscreteSystemImplementationProcessi
ng[rTildeIm, initialConditions, systemParameters,
procedureName];
SequencePlot[rInfoIm[[1;;500]],
SequenceSamplingFrequency→Rs, Joined→True,
```

```
StemPlot→False,AxesLabel→{"Time
(s)","Amplitude"}]
```

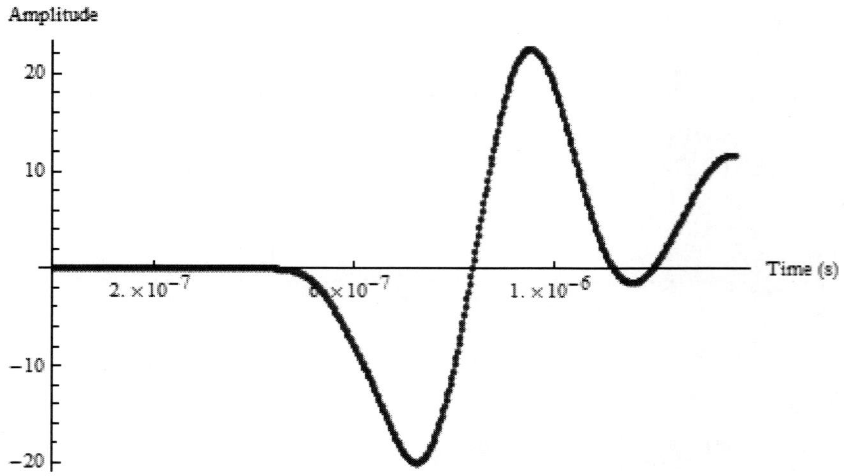

Figure 49. Time response of of real signal at (G).

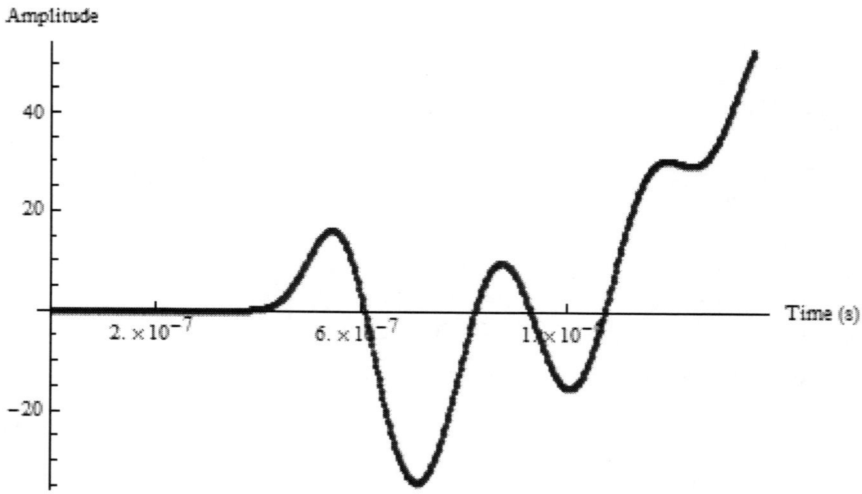

Figure 50. Time response of of imaginary signal at (G).

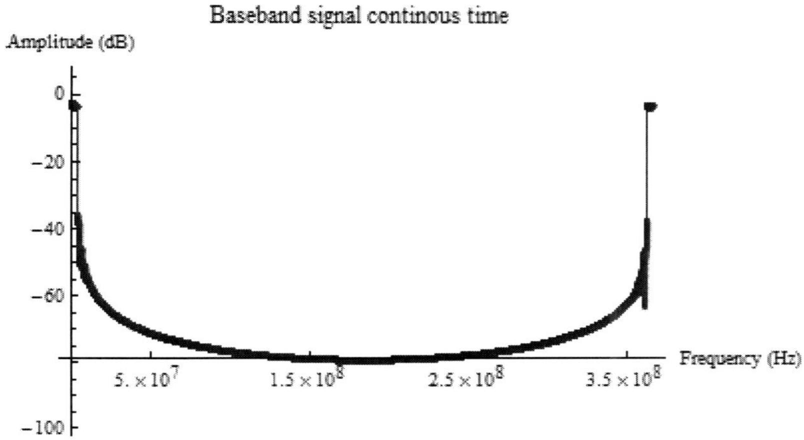

Figure 51. Frequency response of signal at (G).

```
norm=Length[rInfo];
spectrumrInfo=1/norm
SequenceDiscreteFourierTransform[rInfo];
SequencePlot[20 Log[10,Abs[spectrumrInfo]],
Joined→True,
StemPlot→False,
SequenceSamplingFrequency→Tu,PlotLabel→
"Baseband signal continous
time",AxesLabel→{"Frequency (Hz)","Amplitude
(dB)"},PlotRange→{All,{-100,6}}]
```

Next step is sampling.

```
rData=rInfoRe[[1;;Length[t]-1;;20]]
+I*rInfoIm[[1;;Length[t]-1;;20]];
SequencePlot[Re[rData[[1;;20]]], Joined→True,
SequenceSamplingFrequency→1/T,AxesLabel→{"Time
(s)","Amplitude"}]
SequencePlot[Re[rData[[1;;20]]], Joined→False,
SequenceSamplingFrequency→1/T,AxesLabel→{"Time
(s)","Amplitude"}]
```

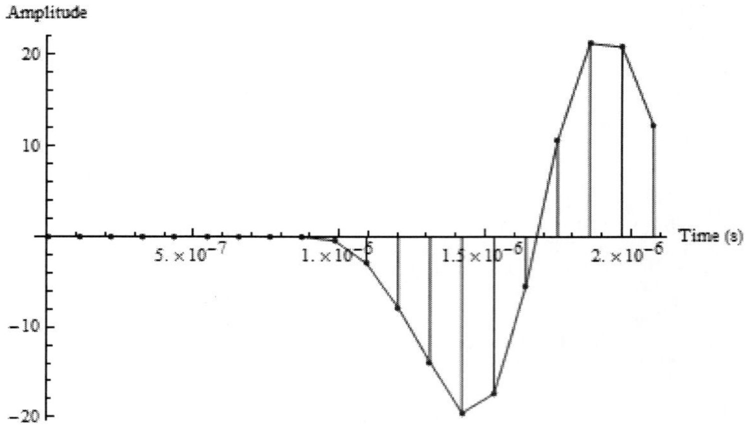

Figure 52. The form of sampled signal at (H).

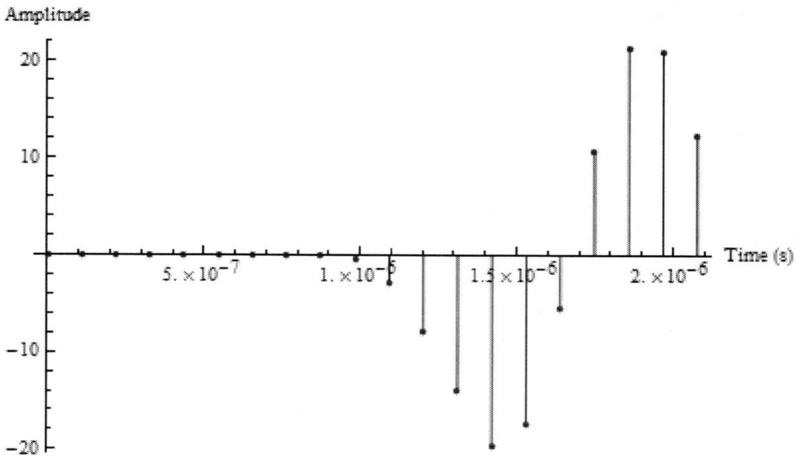

Figure 53. Discret time respons after sampling.

```
norm=Length[rData];
spectrumrData=1/norm SequenceDiscreteFourierTransform[rData];
SequencePlot[20 Log[10,Abs[spectrumrData]], Joined→True, StemPlot→False, SequenceSamplingFrequency→Tu,PlotLabel→"Baseband signal continous time",AxesLabel→{"Frequency (Hz)","Amplitude (dB)"}(*,PlotRange→{All,{-34,6}}*)]
```

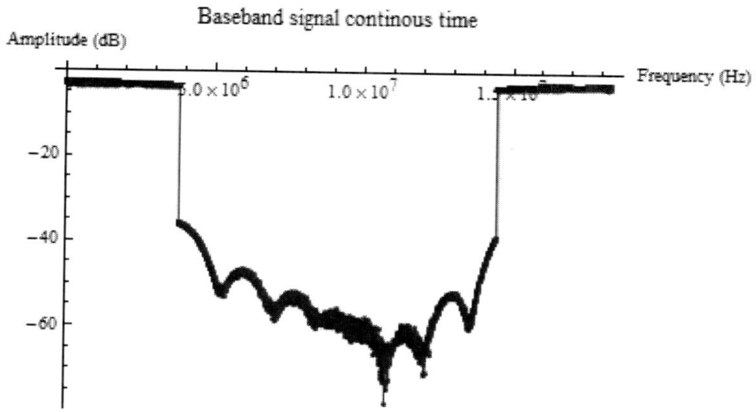

Figure 54. Frequence response at (H).

Last to block are FFT and slicer.

```
info2N=1/FS*Fourier[rData,FourierParameters→{1,-1}];
infoH = ListToSequence[Table[0,{i,1,M}]];
infoH[[1;;M/2]]=info2N[[1;;M/2]];
infoH[[-M/2;;-1]]=info2N[[M/2+1;;M]];
```

Slicer

```
alphabet={-1-I,-1+I,1-I,1+I};
infoHnew=infoH // SequenceToList;
inf=Table[infoHnew[[j]]-alphabet,{j,Length[infoHnew]}];
(*   Q=Ordering[inf/.z_Complex→{Re[z],Im[z]}   //Abs,1] *)
Q=Table[Ordering[inf[[j]]/.z_Complex→{Re[z],Im[z]
} //
Abs,1],{j,Length[infoHnew]}] // SequenceToList;
alphabet=alphabet[[Q[[;;]]]] // ListToSequence;
SequencePlot[Re[alphabet[[1;;20]]], Joined→False,
SequenceSamplingFrequency→2/T,AxesLabel→
{"Time (s)","Amplitude"}]
```

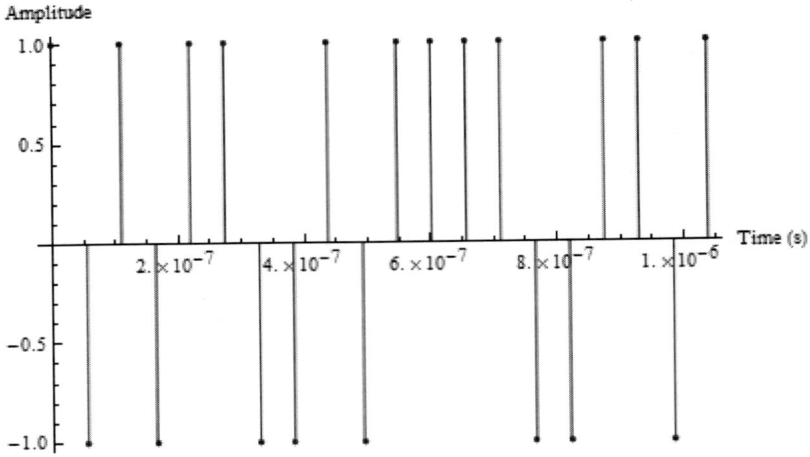

Figure 55. Real decoded signal from discrete time signal at (J).

```
SequencePlot[Im[alphabet[[1;;20]]], Joined→False,
SequenceSamplingFrequency→2/T,AxesLabel→
{"Time (s)","Amplitude"}]
```

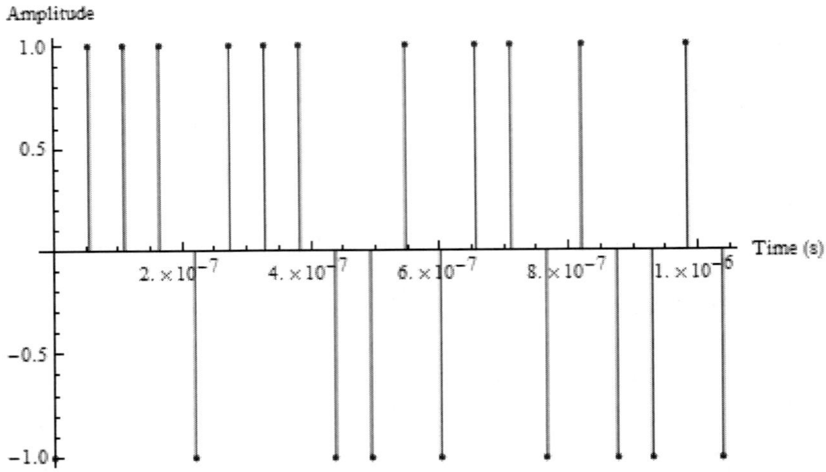

Figure 56. Imaginary decoded signal from discrete time signal at (J).

```
ListPlot[alphabet/.z_Complex→{Re[z],Im[z]},PlotMa
rkers→Automatic,Frame→True]
```

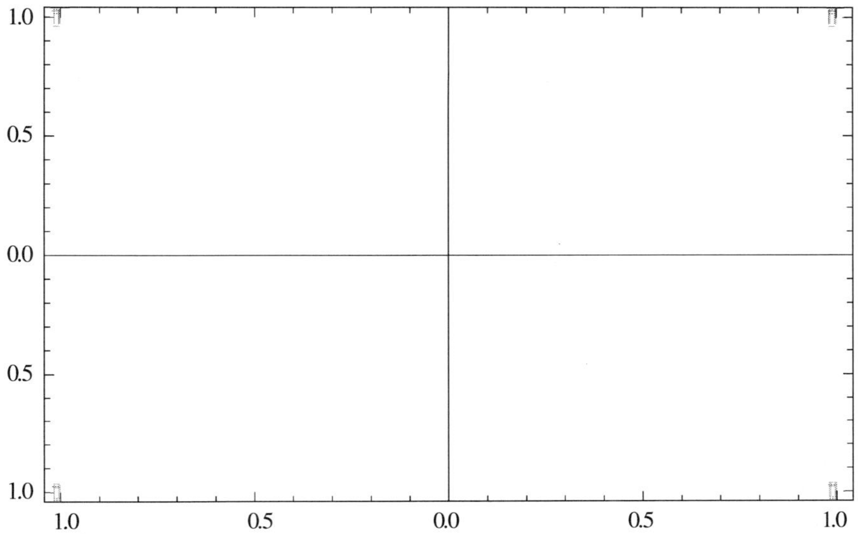

Figure 57. Constellation of decoded signal at (J).

Example 5 - Symbolic Analysis as a Universal Tool for Deriving Properties of Non-Linear Algorithms – Case Study of EM Algorithm

In this section we introduce an original approach and method for analyzing a popular and frequently cited tutorial paper on Expectation-Maximization (EM) algorithm. The original paper has sufficient information to understand the algorithm. Using tools of computer algebra system and methods of Symbolic processing, the typewriting errors are discovered and the re-derived equations are used for automatic generation of algorithmic code. The examples are evaluated using automatically derived code and the final numerical values agree with the values from the original paper. The derived results are used for further optimization, such as deriving the computational error in the closed form. From the closed form solutions, the precision can be derived in terms of number of iterations, or the minimal number of iterations can be expressed in terms of the required precision. This helps to optimize the algorithm parameters and automated-proven properties so that the algorithm becomes more efficient.

The Expectation-Maximization Algorithm

The Expectation-Maximization algorithm is ideally suited to estimate the parameters of a probability function, especially when direct access to all of the data is impossible [18]. A brief description of the algorithm can be found in literature [18] that is suitable for signal processing practitioners. An illustrative example is also presented there. Unfortunately, the algorithmic steps there disagree with the theoretical statements, and do not provide the numerical results given in Table of [18].

Those who want to test the knowledge learned from the example may be confused about the reason for such disagreements. The error can be in the initial statements, in the derived algorithmic steps, or it can be typewriting error.

A computer algebra system (specifically Mathematica) and symbolic processing are used in this paper to start analysis from the theoretical statements, derive the algorithmic code, calculate the final results, and generate error-free derived simulation code.

It is important to emphasize the role of the new analysis of the EM algorithm because EM is the backbone for algorithms that are used in significant analyse, measurement data processing and simulations [1], [7], [8].

The main goal of EM is approximation Maximum Likelihood from incomplete data. The most general form starts from the pdf of the incomplete data:

$$g(y|\theta) = \int_{\chi(y)} f(x|\theta)dx \tag{1}$$

where $f(x|\theta)$ is the probability density function of complete data, and θ is set of parameters. An observation y determines a subset of χ that is denoted as $\chi(y)$.

The EM algorithm consists of two steps: the E-step and the M-step. The E-step (Expectation step) takes an expectation with respect to the unknown variables using the current estimate of the parameter and conditioned upon the observation. The M-step (Maximization step) then provides a new estimation of the parameters, in that it produces maximum-likelihood (ML) estimates of parameters when there is many-to-one mapping. These two steps are iterated until convergence [18].

For the E-step compute:

$$Q(\theta|\theta^{[k]}) = E[\log f(x|\theta)|y,\theta^{[k]}]$$

where $Q(\cdot)$ is $Q-$ function.

For the M-step compute:

$$\theta^{[k+1]} = \arg\max_{\theta} Q(\theta|\theta^{[k]})$$

An Example Problem

The example problem is to find the unknown value of p using the EM algorithm. In this example the initial value of p is defined as a symbol, and we know that its true value is 1/2 (as specified in the example in [34]). The application of EM to this example is calculated symbolically (using Mathematica), and we show mathematical description.

The expressions which are derived in this section, were not derived in [18] and cannot be obtained by hand. On the other side, the estimated parameter p is obtained with the value of 0.52 in [18], but nowhere it explains the deviation from the true value of 0.5. We obtain the exact value of 0.5 and explain the way how we get it.

This section repeats *Example* from [18]. The application of EM to this example is calculated symbolically (using Mathematica), and we show mathematical description. Let a set of random variables is $\{X_1, X_2, \ldots, X_m\}$. For example X_1 represents the number of round dark objects, X_2 represents the number of square dark objects, and X_m represents the number of light objects. Let $x_1, x_2, \ldots, x_m$ be a sequence of outcomes of the random variables $X_1, X_2, \ldots, X_m$ that have been observed. It is assumed that X_i is independent of X_j for $i \neq j$. If the number of outcomes (total number of observed objects) is n, then x_i are nonnegative integers such that

$$\sum_{i=1}^{m} x_i = n$$

If X_1, X_2, ..., X_m are mutually exclusive events with $P(X_1 = x_1) = p_1$, ..., $P(X_m = x_m) = p_m$, then the probability X_1 that occurs x_1 times, ..., X_m occurs x_m times is given by

$$P(X_1 = x_1, X_2 = x_2, \ldots, X_m = x_m) = n! \prod_{i=1}^{m} \frac{p_i^{x_i}}{x_i!}$$

where p_i are constants with $p_i > 0$ and

$$\sum_{i=1}^{m} p_i = 1$$

The joint distribution of X_1, X_2, ..., X_m is a multinomial distribution and $P(X_1 = x_1, X_2 = x_2, \ldots, X_m = x_m)$ is given by the corresponding coefficient of the multinomial series

$$(p_1 + p_2 + \cdots + p_m)^n$$

Assume further that probabilities are also known. As an example, assume that the objects are known to be trinomial distributed, that is $m = 3$, as in [18]. Assume further that probabilities are also known

$$p_1 = \frac{1}{4}$$

$$p_2 = \frac{1}{4} + \frac{p}{4}$$

Then, from (6) follows

$$p_3 = 1 - (p_1 + p_2) = \frac{1}{2} - \frac{p}{4}$$

We can use a notation for probability density function

$$f(x_1, x_2, x_3 \mid p) = P(X_1 = x_1, X_2 = x_2, X_3 = x_3)$$

For the unknown parameter p, the function becomes

$$f(x_1, x_2, x_3 \mid p) = \frac{n!}{x_1! x_2! x_3!} \left(\frac{1}{4}\right)^{x_1} \left(\frac{1}{4} + \frac{p}{4}\right)^{x_2} \left(\frac{1}{2} - \frac{p}{4}\right)^{x_3}$$

Suppose that a feature extractor is employed that can distinguish the color of the objects, but cannot distinguish the shape. From the measurement, the number of dark objects is known, $y_1 = x_1 + x_2$, but the number of x_1 or x_2 is unknown. The basic idea of EM algorithm is to find expressions that can be used for iterated computing of x_1 or x_2 based on known $y_1 = x_1 + x_2$ and distribution function.

Expectation Step

In the E-step, the expected values of x_1, x_2, and x_3, have to be computed using initial estimate and measured data. The algorithm is derived using the knowledge of the multinomial distribution.

Starting with measured y_1 (where $y_1 = x_1 + x_2$) we have to derive expressions that can be used in iterated algorithm in order to find the number of objects x_1 and x_2, and the parameter p. In other words, we have to derive expected values of x_1 and x_2 for measured y_1 and estimate of p in the k th iteration

$$x_1^{[k+1]} = E[x_1 \mid y_1, p^{[k]}]$$

$$x_2^{[k+1]} = E[x_2 \mid y_1, p^{[k]}]$$

The expected values can be derived using the multinomial distribution (assuming that $x_1 + x_2 + x_3 = n$)

$$P(X_1 = x_1, X_2 = x_2, X_3 = x_3) = n! \frac{p_1^{x1}}{x_1!} \frac{p_2^{x2}}{x_2!} \frac{p_3^{x3}}{x_3!}$$

Notice that $(X_1 + X_2, X_3)$ is binomial with class probability $(p_1 + p_2, p_3)$, then (assuming $p_1 + p_2 + p_3 = 1$, $y_1 + x_3 = n$) the probability $P(Y_1 = y_1, X_3 = x_3)$ becomes

$$P(Y_1 = y_1, X_3 = x_3) = n! \frac{(p_1 + p_2)^{y1}}{y_1!} \frac{p_3^{x3}}{x_3!}$$

Assume that the probability of some event x is known, say $P(x)$, and the number of events can be from 0 to n, the expectation value of x is defined by

$$E[x] = \sum_{x=0}^{n} x\, P(x)$$

Noticing that $y_1 = x_1 + x_2$ and $0 \le x_1 \le y_1$, expectation of x_1 requires to know conditional probability $P(x_1)$,

$$E[x_1] = \sum_{x=0}^{y_1} x_1\, P(x_1)$$

Assuming that y_1 has occurred, the conditional probability $P(x_1)$ equals

$$P(x_1) = \frac{P(X_1 = x_1, Y_1 = y_1)}{P(Y_1 = y_1)}$$

Applying those definitions on $x_1^{[k+1]}$, the iterated algorithm can be derived. Unfortunately, the presented relations in [18] disagree with expected.

Maximization Step

The maximum of some function, say $f(x_1, x_2, x_3 \mid p)$, can be obtained by taking the derivative of the function f with respect to parameter p, equating to zero, and solving for p. Since the logarithm is monotonically increasing, maximizing $\log(f)$ is equivalent to maximizing f.

The function f is known in terms of the parameter p, and we can try to solve for p the following equation

$$\frac{\mathrm{d}}{\mathrm{d}p} \log\big(f(x_1, x_2, x_3 \mid p)\big) = 0$$

If solution exists, it becomes the second part of EM algorithm.

Implementation of Computer Algebra System Code into EM Algorithm

We re-derive the expressions obtained in [18] because they do not produce the expected results in the numerical example presented in [18]. The main motive is to avoid manual derivation and thus derive all relations without error. All derivations are in a document called notebook that looks like technical paper in electronic form. There are cells with textual description, such as title, section head, text with formula, but also cells named input and output. In the input cells we write expressions that can be evaluated.

Entering Knowledge in CAS

The simplest way to re-derive equations is to start with definitions. The first line of the first input cell can be the definition of the trinomial distribution

```
f[x1_, x2_, x3_, p1_, p2_, p3_] := (x1 + x2 + x3) ! p1^x1/x1 ! p2^x2/x2 ! p3^x3/x3 !
```

By executing the input cell there is no visible output cell. As a matter of fact, we define new knowledge that becomes a part of the computer algebra system. The meaning of the code is that we define a function called f, a function with 6 arguments, (x1, x2, x3, p1, p2, p3), where instead of any argument with _ we can use any other symbol or number (instead of x1_ we can use x_1). The symbol := tells to computer algebra systems to remember the expression that follows, but without execution, and so there is no output cell.

Once the knowledge is inputted to computer algebra systems, we can use for presentation or evaluation. The mathematical expressions may look different than a code. Let us replace symbolic arguments with symbols that appear in textbooks (use x_1 instead of x1), and present in the traditional form (transform the expression using command //TraditionalForm).

This expression still does not look same as equation $P(X_1 = x_1, X_2 = x_2, X_3 = x_3) = n!\frac{p_1^{x_1}}{x_1!}\frac{p_2^{x_2}}{x_2!}\frac{p_3^{x_3}}{x_3!}$). Therefore, we evaluate the expression again for the same symbolic arguments, and in the next line use the expression (% stays for previous expression) and apply substitution as in equation $\sum_{i=1}^{m} x_i = n$, and finally, display the result in the traditional form

```
In[2]:= f[x1, x2, x3, p1, p2, p3] // TraditionalForm
```

Out[2]//TraditionalForm=

$$\frac{(x_1 + x_2 + x_3)!\, p_1^{x1}\, p_2^{x2}\, p_3^{x3}}{x_1!\, x_2!\, x_3!}$$

Now, we obtain the same result as equation $P(X_1 = x_1, X_2 = x_2, X_3 = x_3) = n!\frac{p_1^{x1}}{x_1!}\frac{p_2^{x2}}{x_2!}\frac{p_3^{x3}}{x_3!}$. This means that inputted knowledge is visually identical as in textbook. Notice that the last output is a result of the line that follows input In[3] that is actually input In[4] and the displayed output is Out[4].

The binomial distribution can be defined in a similar way

```
In[3]:= f[x1, x2, x3, p1, p2, p3];
        % /. x1 + x2 + x3 → n // TraditionalForm
```

Out[4]//TraditionalForm=

$$\frac{n!\, p_1^{x1}\, p_2^{x2}\, p_3^{x3}}{x_1!\, x_2!\, x_3!}$$

This defines a function called f, a function with 5 arguments, (y1, y2, p1, p2, n). The head of the function is the same as for the trinomial distribution, but the difference is in the number of arguments. Instead of any argument with _ we can use any other symbol or number (instead of y2_ we can use x_3, and instead of p1_ we can use $p_1 + p_2$). Replacing arguments with appropriate symbols, and using the command for displaying expression in the traditional form, equation $P(Y_1 = y_1, X_3 = x_3) = n!\frac{(p_1 + p_2)^{y1}}{y_1!}\frac{p_3^{x3}}{x_3!}$ is derived

```
In[6]:= f[y1, x3, p1 + p2, p3, n] // TraditionalForm
```

Out[6]//TraditionalForm=

$$\frac{n!\, p_3^{x3}\, (p_1 + p_2)^{y1}}{x_3!\, y_1!}$$

Derivation of the Expectation Step

In order to find out expectation of x_1, the conditional probability $P(x_1)$ should be computed using equation $P(x_1) = \dfrac{P(X_1 = x_1, Y_1 = y_1)}{P(Y_1 = y_1)}$

```
In[7]:= cp1 = f[x1, x2, x3, p1, p2, p3] / f[x1 + x2, x3, p1 + p2, p3, x1 + x2 + x3] ;
        Px1 = cp1 /. x2 → y1 - x1 // Simplify;
        % /. {p1 → p₁, p2 → p₂, x1 → x₁, y1 → y₁};
        % // TraditionalForm
```

Out[10]//TraditionalForm=

$$\frac{y_1!\, p_1^{x_1} (p_1 + p_2)^{-y_1} p_2^{y_1 - x_1}}{x_1!\,(y_1 - x_1)!}$$

In the nominator the trinomial distribution is used, while in the denominator the binomial distribution is applied. The command Simplify is use to find the simplest form of the expression. Next, the derived expression is used in equation $E[x_1] = \sum_{x=0}^{y_1} x_1 P(x_1)$, and the final result is not the same as in [18]

```
In[11]:= ex1 = Sum[x1 Px1, {x1, 0, y1}];
         % /. {p1 → p₁, p2 → p₂, y1 → y₁}
         % // TraditionalForm
```

Out[12]=

$$\frac{p_1 y_1}{p_1 + p_2}$$

Out[13]//TraditionalForm=

$$\frac{p_1 y_1}{p_1 + p_2}$$

The same procedure is used for deriving expected value of x_2, but this time the same result is as in [18]

In[14]:= $\texttt{cp2} = \frac{\texttt{f[x2, x1, x3, p2, p1, p3]}}{\texttt{f[x1 + x2, x3, p1 + p2, p3, x1 + x2 + x3]}};$

```
Px2 = cp2 /. x1 → y1 - x2 // Simplify;
```

In[16]:= $\texttt{ex2} = \sum_{\texttt{x2=0}}^{\texttt{y1}} \texttt{x2 Px2};$

```
% /. {p1 → p₁, p2 → p₂, y1 → y₁};
% // TraditionalForm
```

Out[18]//TraditionalForm=

$$\frac{p_2\, y_1}{p_1 + p_2}$$

Sum of expected values of x_1 and x_2 should be y_1, that is obvious from the derived expressions using computer algebra systems, but also proves that one of two results in [18] is not correct. Now, it is obviously that the sign – in the denominator of the expectation of x_1 is not correct in [18]. Also, in the original text [18], it used y and y_1 for the same variable, that may be also confusing for some readers.

The presented derivation in this section shows that all results can be found out automatically, and intermediate results and defined functions can be visually identified to be the same with those from textbooks.

Once the general expressions are derived, a particular solution follows as a special case. For example, if the probabilities are known, they can be denoted by subscript e

In[19]:= $p_{1e} = \frac{1}{4};$

$$p_{2e} = \frac{1}{4} + \frac{p}{4};$$

The expected value of x_1 for this specific case can be derived by substituting parameters in the general solution for x_1

```
In[21]:= ex1 /. {p1 → p1e, p2 → p2e, y1 → y1};
         % // TraditionalForm
```

Out[22]//TraditionalForm=

$$\frac{y_1}{4\left(\frac{p}{4}+\frac{1}{2}\right)}$$

Following the same procedure, the expected value of x_2 is derived

```
In[23]:= ex2 /. {p1 → p1e, p2 → p2e, y1 → y1};
         % // TraditionalForm
```

Out[24]//TraditionalForm=

$$\frac{\left(\frac{p}{4}+\frac{1}{4}\right)y_1}{\frac{p}{4}+\frac{1}{2}}$$

The derived results differ from that in [18], and instead of p, the automatically derived expressions have $p/2$.

Derivation of the Maximization Step

For known probabilities of a specific case, the trinomial distribution can be expressed in terms of x_1, x_2, x_3, and the parameter p

```
In[25]:= f[x1, x2, x3, p1, p2, p3];
         % /. {p1 → p1e, p2 → p2e, p3 → (1 - p1e - p2e)};
         fe = % /. {x1 → x1, x2 → x2, x3 → x3};
         % /. x1 + x2 + x3 → n;
         % // TraditionalForm
```

Out[29]//TraditionalForm=

$$\frac{4^{-x_1}\, n!\left(\frac{1}{2}-\frac{p}{4}\right)^{x_3}\left(\frac{p}{4}+\frac{1}{4}\right)^{x_2}}{x_1!\, x_2!\, x_3!}$$

The same expression is presented in [18]. Since this is a continuous function in p, the first derivative can be computed

```
In[30]:= d = D[Log[fe], p] // Simplify
```

Out[30]= $$\frac{(-2+p)\, x_2 + (1+p)\, x_3}{(-2+p)\,(1+p)}$$

The parameter p can be expressed in terms of for the extreme (and again, the same expression is presented in [18])

```
In[31]:= sol1 = Solve[d == 0, p] // Flatten
```

Out[31]= $$\left\{p \to \frac{2\, x_2 - x_3}{x_2 + x_3}\right\}$$

But, the value of x_2 is unknown, Instead of x_2, we can use the expected value of x_2, that was derived in the expectation step. In order to obtain the same form of the final results as in [18], some rearrangements can be evaluated (to collect all terms in numerator and denominator that are multiplied by p)

```
In[32]:= sol1 /. x2 → ex2;
         s = % /. {p1 → p1e, p2 → p2e, y1 → y1} // Simplify
         pnew = Collect[Numerator[p /. s], p] / Collect[Denominator[p /. s], p]
```

Out[33]= $$\left\{p \to \frac{-(2+p)\, x_3 + 2\,(1+p)\, y_1}{(2+p)\, x_3 + (1+p)\, y_1}\right\}$$

Out[34]= $$\frac{-2\, x_3 + 2\, y_1 + p\,(-x_3 + 2\, y_1)}{2\, x_3 + y_1 + p\,(x_3 + y_1)}$$

The final result is not the same as in [18]. It seems that $2p$ is used instead of p in the recurrence relation in [18].

Derivation of the Code

Once the expressions are derived from inputting the knowledge into computer algebra systems, the algorithm in the form of code can be simply transformed. The code can be derived for some input parameters, such as the total number of objects, number of dark or light objects, initial guess for p, required accuracy, or probabilities

```
In[35]:= Clear[k, p]
         n = 100;
         y1 = 63;
         x3 = n - y1;
         p[0] = 0;
         a = 10^-7;
         p1e = 1/4;
         p2e = 1/4 + p/4;
```

The expressions used in the code is a special case of the general expressions in terms of the known parameters or the previous value of p

```
In[43]:= pnew = pnew /. {x3 → x3, y1 → y1} /. p → p[k - 1];
         x1new = ex1 /. {p1 → p1e, p2 → p2e} /. p → p[k - 1];
         x2new = ex2 /. {p1 → p1e, p2 → p2e} /. p → p[k - 1];
```

Iterated algorithm can be a while loop, in that new number ob objects or new p is calculated

```
In[46]:= e = 1; k = 1;
         While[e > a, {
            p[k] = pnew;
            x1[k] = x1new;
            x2[k] = x2new;
            e = p[k] - p[k - 1];
            k = k + 1}]
```

Finally, the results of iterations can be viewed as table of values

```
Table[N[{i, x1[i], x2[i], p[i]}, 8], {i, 1, k - 1}] // TableForm
```

Number of displayed digits can be also specified

Out[48]//TableForm=

1.0000000	31.500000	31.500000	0.37956204
2.0000000	26.475460	36.524540	0.49029997
3.0000000	25.298157	37.701843	0.51409288
4.0000000	25.058740	37.941260	0.51883995
5.0000000	25.011514	37.988486	0.51977276
6.0000000	25.002255	37.997745	0.51995551
7.0000000	25.000441	37.999559	0.51999129
8.0000000	25.000086	37.999914	0.51999829
9.0000000	25.000017	37.999983	0.51999967
10.000000	25.000003	37.999997	0.51999993
11.000000	25.000001	37.999999	0.51999999

The values from table are the same as in [18]. Therefore, using computer algebra systems, we derive all expressions and code for EM algorithm. Also, we avoid typewriting errors in the original text, prove all of the steps, and compute the numerical results for a specific case. For example, it is obvious that y_1 cannot be 100 [18], because $n = 100$ and $y_1 < n$. A typewriting error is in the derivation of expectation step [18] where index in the summation formula is from 10. Another typewriting error is in the derivation of expectation step [18] where x_2 should be used instead of x_1.

Analytical Solution of the EM Algorithm

The EM algorithm can be further analyzed by employing the symbolic expressions derived using computer algebra systems. Instead of the numerical analysis, a closed form expressions can be derived. This can help us to gain insight into how some system works and understand the properties of the algorithm.

Finding Closed Form Solution

Firstly, we generate an equation that describes iterated computing of parameter p, knowing the initial value

```
In[49]:= Clear[k, p]

In[50]:= eq1 = {p[k + 1] == (pnew /. p → p[k]), p[0] == 0}
```

$$\text{Out[50]=}\ \left\{p[1+k] == \frac{-2\,x_3 + 2\,y_1 + p[k]\,(-x_3 + 2\,y_1)}{2\,x_3 + y_1 + p[k]\,(x_3 + y_1)},\ p[0] == 0\right\}$$

Command RSolve can be used to derive a closed form expression for p in terms of the iteration step k and other parameters kept as symbols

```
In[51]:= sol2 = RSolve[eq1, p[k], k] // Flatten // FullSimplify
```

$$\text{Out[51]=}\ \left\{p[k] \to -\frac{2\,(x_3 - y_1)\left(\left(-1 - \frac{x_3}{y_1}\right)^k - \left(-3 - \frac{3\,y_1}{x_3}\right)^k\right)}{2\left(-1 - \frac{x_3}{y_1}\right)^k (x_3 - y_1) - (x_3 + y_1)\left(-3 - \frac{3\,y_1}{x_3}\right)^k}\right\}$$

In the specific case, when numerical values of some parameters are known, the closed form expression can be in terms of k, only

```
In[52]:= sol3 = sol2 /. {x3 → x3, y1 → y1} // Simplify
```

$$\text{Out[52]=}\ \left\{p[k] \to -\frac{13\left(\left(\frac{100}{63}\right)^k - \left(\frac{300}{37}\right)^k\right)}{13\left(\frac{100}{63}\right)^k + \left(\frac{12}{37}\right)^k 25^{1+k}}\right\}$$

The numerical values of p in each iteration step can be displayed in table format

```
In[53]:= Table[N[p[k] /. sol3, 8], {k, 1, 11}] // TableForm

Out[53]//TableForm=
        0.37956204
        0.49029997
        0.51409288
        0.51883995
        0.51977276
        0.51995551
        0.51999129
        0.51999829
        0.51999967
        0.51999993
        0.51999999
```

Closed form expression can be used to find out the exact value after infinite number of iteration

```
In[54]:= Limit[p[k] /. sol3, k → +Infinity]
         % // N
```

$$\text{Out[54]}= \frac{13}{25}$$

Out[55]= 0.52

Any of the parameters from the general solution can be changed to compute another set of values, such as for any value of n and specified exact value for p (named p_{True})

```
In[56]:= Clear[n]
```

$$p_{True} = \frac{1}{2};$$

$$p_{1e} = \frac{1}{4}\ /.\ p \to p_{True};$$

$$p_{2e} = \frac{1}{4} + \frac{p}{4}\ /.\ p \to p_{True};$$

```
y1 = Round[p1e n] + Round[p2e n];
x3 = n - y1;
Limit[p[k] /. sol2 /. n → 100, k → +Infinity]
% // N
```

$$\text{Out[62]}= \frac{13}{25}$$

Out[63]= 0.52

Since we already have a limit value as closed form solution in terms of n, the exact values can be computed for a range of different n

```
nrange = Range[50, 150];
pLimit = Limit[p[k] /. sol2 /. n → nrange, k → Infinity];
ListPlot[Transpose[{nrange, pLimit}],
 Filling → Axis, AxesOrigin → {50, 0.45}, AxesLabel → {n, p}]
```

Those results can be illustrated as it is shown in Figure 58. The probability can be from the range 0.45 to 0.55, depending on n. Regardless of the fact that initial value of p is 1/2, named p_{True} in the code, the computed value of is from the range 0.45 to 0.55 as exact result of the algorithm after infinite number of iterations. The exact value $p=0.5$ can be obtained for a large number of objects, or as a mean value for several n.

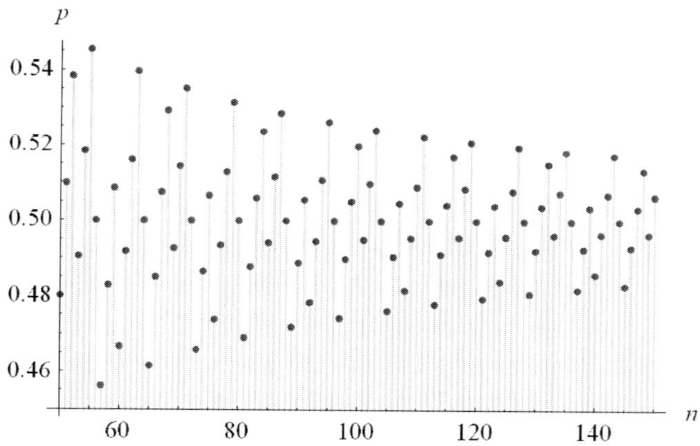

Figure 58. Estimate value of p in terms of the number of observed objects n.

Finding Number of Iterations K

Usually, the errors in iterated algorithms are estimated. Since we already have the exact value after infinite number of iterations, and the current value at any iteration, the error can be expressed in closed form

```
error = (Limit[p[k] /. sol2 /. n → 100, k → +Infinity] -
      p[k] /. sol2 /. n → 100) // Simplify
```

$$\frac{494}{325+625\left(\frac{189}{37}\right)^k}$$

$$a = \frac{1}{2^b}$$

$$2^{-b}$$

In addition, we specify the acceptable error as $a = 1/2^b$, where b is the number of accurate bits. Solving the equation when error is equal to required number of bits, we can determine b in terms of the iteration index k

```
sol4 = Solve[error == a, b] // Flatten // Simplify
```

$$\left\{b \to -\frac{\text{Log}\left[\frac{494}{325+625\left(\frac{189}{37}\right)^k}\right]}{\text{Log}[2]}\right\}$$

For tree different number of iteration, it follows that 4 iterations are sufficient for 8 bits,

```
b /. sol4 /. k → {3, 6, 10} // N
b /. sol4 /. k → {4, 7, 11} // N
```

{7.40333, 14.4561, 23.8672}

{9.7516, 16.8089, 26.22}

Figure 59 shows the number of accurate bits is liner function of number of iterations

```
Plot[b /. sol4, {k, 1, 11},
 PlotStyle → Thick, AxesLabel → {k, b},
 GridLines → {{4, 7, 11}, {8, 16, 24}},
 Ticks → {{0, 2, 4, 7, 9, 11}, {0, 8, 16, 24}}]
```

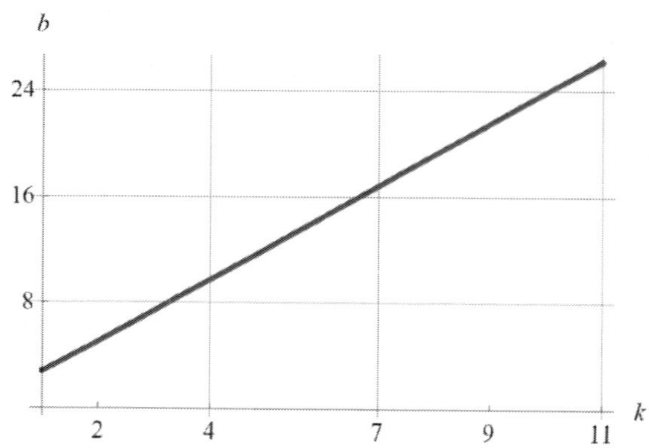

Figure 59. Number of exact bits of estimated p vs. number iterations k.

Instead of number of accurate bits, we can set number of accurate decimal digits

$$a = \frac{1}{10^{digits}}$$

$$10^{-digits}$$

Solving the equation when error is equal to required number of accurate decimal digits, we can determine minimal number of iterations

```
sol5 = Solve[error == a, k] // Flatten // Simplify
```

$$\left\{k \to \text{Log}\left[-\frac{13}{25} + 247 \times 2^{1+digits}\, 5^{-4+digits}\right] \Big/ \text{Log}\left[\frac{189}{37}\right]\right\}$$

The minimal number of iterations is a function of digits

```
k /. sol5 /. digits → Range[0, 15, 3] // N
```

```
{-0.801956, 4.0911, 8.32723, 12.563, 16.7987, 21.0344}
```

For a 6-digit precision, the minimal number of iterations is 9

```
Ceiling[k /. sol5 /. digits → 6 // N]
```

```
9
```

Recalling the table with results, we can see that the 9th iteration has 6-digit precision.

This result can be used for fixed number of iterations instead of while loop. This is more suitable for numerical algorithms, especially when cash instructions can improve the speed of computing (no branch at the end of set of instructions).

In a similar way, by increasing n, the maximal number can be determined for the required precision. For n from 20 to 10000, 9th iteration is sufficient for 6-digit precision.

IV. The Importance of Methods of Computer Algebra System and Symbolic Processing and Future Applications

Contemporary trends to use very sophisticated algorithms combine expertise in many areas, such as multirate system design, control engineering, analog electronic circuit VLSI design, and signal processing. This trend caused programming to become a task of knowledge accumulation and an efficient human-machine interface. The computer algebra systems allow analysis and processing on the natural way and so lacking in creativity of the work computer systems do. Symbolic analysis and processing are accurate and are not subject to errors. The computer algebra systems provide a simple graphical interface, so the many complex operations can be performed by activating the graphic icons. Block diagram of the system is not just a still image, but it contains all the necessary elements for the generation of high quality schemes for addressing, processing, and simulation. The focus of the problem is not the analysis of numerical data or multiple repetitions of numerical experiments, than the setting of problem and preparing the data, so the result of analysis and signal processing obtains the desired shape. Superiority of symbolic computation against numerical computation was shown: (a) by the example system with an algebra loop, computer algebra systems yielded the exact solution while the traditional numerical approach failed, (b) by the closed-form solution of a nonlinear LMS subsystem, and (c) by deriving the analytic expression for the error of the iteration. Mixed symbolic-numerical techniques were used for the efficient and effective search for optimal solutions: (a) high-speed digital filter and (b) programmable logic DSP system based on minimum-Q elliptic function. It was pointed out that those who use powerful computer algebra systems had to thoroughly understand the assumptions that underlie the software and the limits of computer algebra systems: (a) analytic computation of the z-transform might fail, (b) the closed-form response of a non-linear system might not be found, and (c) the symbolic matrix analysis with a large number of symbols might become impractical. Benefits of symbolic methods and the role of computer algebra systems were highlighted from the viewpoint of both academia (derivation of time and frequency response, proving system properties) and industry (LMS algorithm, noise analysis, verification of realizations, design alternatives in multirate systems, CD-media controller, and pitch controller). Apart from the main advantage of the exact arithmetic that computer algebra systems have, they also have other advantages such as: interactivity–we give questions and directly get answers, arbitrary precision-arithmetic or even exact arithmetic can be used to utilize some of the direct (but numerically error proned) methods efficiently in control system design–apart of the multiprecision arithmetic we may use, we can select specific precision for work, visualization of results–we can view in 2-D and 3-D graphics results (XY plots, XYZ plots, polar plots, log plots etc.) and import/export results to known graphics formats, the existence of hundreds of built-in procedures covering both general and specialized areas of mathematics such as Laplace transforms, Inverse Laplace transforms, Z-transforms, inverse Z-transforms, Smith form of polynomial matrices, the existence of unique high-level programming languages and programming environments allowing specific procedures to be written, they are usually available for different types of computer platforms, computer algebra system enables us to fully participate in the mathematical exploration, by exploring different scenarios by asking or being asked ''what-if'' type of questions. The computer algebra systems help us to work with commutative or noncommutative algebra problems, where numerical environments are not suitable. The computer algebra systems can also be used in research to test conjectures — supporting or refute conjectures, to help us to carry out the solutions steps of a mathematical algorithm and avoiding the handcalculation mistakes, to help us to design and build a computer algebra systems for a specialized task in our research area, to help us to customize and improve our algorithms for the solution of a problem, to give us closed form solutions, and thus provide deeper insight to the problem - we can find for example the family of stabilizing compensators of a system in a general parametric framework by retaining the extra degrees of freedom as symbols appearing as free parameters in the expression for the compensator. These free parameters can be then optimized to achieve specific goals associated with the internal stability and/or robustness of the resulting closed loop system, to assist in creating large mathematical tables such as integral tables, tables of special functions and so on, to help us to explore new algorithms and techniques, to give us time to focus on ideas and not on calculations - by helping us for example to theoretically analyze the structural properties of linear and nonlinear systems when some parameters are included in the model and we are looking for generic results instead of particular solutions for specific values of them. However, computer algebra system as part of their large memory requirements and relative poor performance are also accompanied by disadvantages such as difficulty to define the domain of the solution we are looking for, contain particularities that are learned only with experience, specialized areas are not covered by existing general purpose computer algebra system, difficulty to give exact answers in problems where a closed form solution does not exist, e.g., exact solution of a general fifth order polynomial equation or a closed formula solution to a non-linear problem, difficulty to connect with other applications, difficulty to handle large scale problems due to the their excessive computational resources requirement, i.e. difficulty to handle large size of polynomials and polynomial matrices with high degrees and/or large dimensions, symbolic matrix analysis with a large number of symbols may become impractical, may give meaningless generic answers.

REFERENCES

[1] M. Lutovac, “Signal processing and analysis of telecommunications system using computer algebraic system”, Proc. XXIII Simposium of a new technologies in post office and telecommunication traffic - *PosTel* 2005, Belgrade, December 2005.

[2] D. V. Tošić, M. D. Lutovac, “Advances in symbolic simulation of systems,” *The IPSI BgD Transactions on Advanced Research,* vol, 3, no. 1, Jan. 2007.

[3] Gielen, G., Wambacq, P., Sansen, W., “Symbolic analysis methods and applications for analog circuits: A tutorial overview,” *Proceedings of the IEEE,* 1994, pp. 286-304.

[4] Lin, Pen-Min, “Symbolic network analysis,” *Elsevier,* Amsterdam, The Netherlands, EU, 1991.

[5] Wolfram, S., The Mathematica Book, 5th Edition, *Cambridge University Press,* Cambridge, UK, 2003.

[6] MATLAB Version 7, MathWorks, Inc., Natick, MA, USA, 2005.

[7] Lin, Pen-Min, “Symbolic network analysis,” *Elsevier,* Amsterdam, The Netherlands, EU, 1991.

[8] Huelsman, L., Gielen, G., “Symbolic Analysis of analog circuits: Techniques and applications,” *Kluwer Academic Publishers,* Norwell, MA, USA, 1993.

[9] Gielen, G., Wambacq, P., Sansen, W., “Symbolic analysis methods and applications for analog circuits: A tutorial overview,” *Proceedings of the IEEE,* 1994, pp. 286-304.

[10] Fernández, F., Rodríguez-Vázquez, A., Huertas, J., Gielen, G., “Symbolic analysis techniques: Applications to analog design automation,” *Wiley-IEEE Press,* 1997.

[11] Lutovac, D., Tosic, D., Evans, B., "Filter Design for Signal Processing using MATLAB and Mathematica," *Prentice Hall,* Upper Saddle River, NJ, USA, 2001.

[12] Qin, Z., Tan, S., Cheng, C., "Symbolic analysis and reduction of VLSI circuits," *Springer,* 2004.

[13] Riddle, A., Dick, S., "Applied electronic engineering with Mathematica," Addison-Wesley, Reading, MA, USA, 1995.

[14] M. Lutovac, D. Tosic, "Symbolic analysis and design of control systems using Mathematica", *International Journal of Control, Special Issue on Symbolic Computing in Control,* vol.79, no.11, 2006, pp.1368–1381

[15] V. Mladenovic, M. Lutovac, D. Porrat, "Symbolic Analysis as Universal Tool for Deriving Properties of Non-linear Algorithms –Case study of EM Algorithm", *Acta Polytechnica Hungarica* 2013

[16] M. Lutovac, J. Ćertić, Lj. Milić, "Digital Filter Design Using Computer Algebra Systems," *Circuits Syst. Signal Process.,* vol. 29, no. 1, 2010, pp. 51-64.

[17] M. Lutovac, D. Tosic, "Symbolic signal processing and system analysis", *Facta Universitatis, Electronics and Energetics,* vol. 16, no. 3, 2003, pp. 423-431.

[18] T. Moon, "The expectation-maximization algorithm", *IEEE Signal Processing Mag.,* 1997, pp. 47–60.

[19] M. Lutovac, V. Mladenovic, "Development of aeronautical communication system using computer algebra systems", Proc. XXV Simposium of a new technologies in post office and telecommunication traffic - *PosTel* 2007, Belgrade, December 2007.

[20] M. Lutovac, D. Tosic, "High-Speed Filter Design using Mathematica", in *Int. Conf. EUROCON,* 2005, pp. 1626–1629.

[21] H. Johansson and L. Wanhammar, "High-speed recursive filter structures composed of identical all-pass subfilters for interpolation, decimation, and QMF banks with perfect magnitude reconstruction," *IEEE Trans. Circuits and Systems*—II, vol. 46, pp. 16–28, Jan. 1999.

[22] Lj. Milic and M. Lutovac, "High speed IIR filters for QMF banks," in *Proc. Int. Conference TELSIKS, Niš,* 2001, pp. 171-174.

[23] M. Lutovac, D. Tosic, "SchematicSolver Version 2.2", 2010. Available: http://books.google.com/books?id=9ue-uVG__JsC

[24] V. Mladenovic, M. Lutovac, Broadband VHF aeronautical communication system, 15th TELECOMMUNICATIONS FORUM, TELFOR 2007, *IEEE COM SOC Chapter Serbia & Montenegro Belgrade,* Serbia 20, 21 and 22. November 2007, RC#3.16

[25] M. D. Lutovac, V. M. Mladenovic, “Development of aeronautical communication system using computer algebra systems,” in Proc. XXV Simposium of a new technologies in post office and telecommunication traffic - *PosTel* 2007, Belgrade, December 2007.

[26] H. Schulze and C. Luders, Theory and Applications of OFDM and CDMA - Wideband Wireless Communications. *West Sussex: John Wiley & Sons,* 2005.

[27] D. Porrat, “Information Theory of Wideband Communications,” *IEEE Communications Surveys and Tutorials,* vol. 9, no. 2, pp. 2-16, Second Quarter 2007.

[28] R.V. Nee and R. Prasad, OFDM Wireless Multimedia Communications. Norwood, MA: *Artech House,* 2000.

[29] G. Acosta. “OFDM Simulation Using Matlab.” Smart Antenna Research Laboratory. Available: http://www.ece.gatech.edu/research/labs/sarl/tutorials/OFDM/Tutorial_web.pdf 2001.

[30] J. A. Fessler and A. O. Hero, Space-alternating generalized expectationmaximization algorithm, *IEEE Trans. Signal Processing,* vol. 42, pp. 2664–2677, Oct. 1994.

[31] M. Lutovac, D. Tošić, “Symbolic signal processing and system analysis”, *Facta Universitatis, Electronics and Energetics,* vol. 16, no. 3, 2003, pp. 423-431.

[32] B. H. Fleury, D. Dahlhaus, R. Heddergott, and M. Tschudin, “Wideband angle of arrival estimation using the SAGE algorithm”, Proc. IEEE Fourth Int. Symp. Spread Spectrum Techniques and Applications (ISSSTA ’96), Mainz, Germany, Sept. 1996, pp. 79–85.

[33] D. Porrat, “The Diffuse Multipath Component and Multipath Component Visibility”, The 5th European Conference on Antennas and Propagation, to be held at EUR Congressi in Rome, Italy, EuCAP 2011., on 11-15 April 2011

[34] M. D. Lutovac, V. M. Mladenovic, “Post-processing with advanced symbolic simulation and design using SchematicSolver and Mathematica”, 9th International Conference on Telecommunications in Modern Satellite, Cable and Broadcasting Services- TELSIKS – IEEE societies: MTT, AP, Communications, and Region 8., October 7 - 9, 2009. Nis, Serbia&Montenegro.

In: Telecommunications
Editor: Joseph P. Barringer
ISBN: 978-1-63117-141-3

Chapter 3

EXPLICIT FORM OF THREE NOVEL CLASSES OF MULTIPLIERLESS LINEAR PHASE SELECTIVE 2D FIR FILTER FUNCTION

***Vlastimir D. Pavlović*[1,¥] *and Nebojša S. Dončov*[1,€]**
[1]Faculty of Electronic Engineering, University of Niš,
Niš, Republic of Serbia

ABSTRACT

The application of the extremal property of Christoffel-Darboux formula for classical continual orthonormal polynomials to the theory of synthesis of continual filter functions and digital multiplierless linear phase one-dimensional (1D) and two-dimensional (2D) FIR filter functions is completely studied and described by analytical methods, and the results are given in compact explicit form in literature [1-13].

By correct mapping from the continual domain to the z domain without errors and without residues, an ultra selective spike-like pass-band is obtained and unwanted parasitic Gibss effect is completely suppressed. In this chapter, we propose in explicit form three novel classes of multiplierless linear phase selective 2D FIR filter function, of symmetric order $N*N$, obtained directly by using analytical method in z domain. This paper represents introduction in ATLAS 2D FIR filter

¥ E-mail: vlastimir.pavlovic@elfak.ni.ac.rs.
€ E-mail: nebojsa.doncov@elfak.ni.ac.rs.

functions. Three novel classes of 2D FIR filter functions $H_i(N, z_1, z_2)$, $i \in (1,2,3)$ are generated by using cascade realization and by the derivation of two free real integer parameters the following form is obtained:

$$H_i(N, K, z_1, z_2) = [H_i(N-1, z_1, z_2) * H_i(N, z_1, z_2) * H_i(N+1, z_1, z_2)]^K, \quad i \in (1,2,3)$$

The examples of proposed filter functions are considered in the paper and the detailed analysis of amplitude response characteristics, magnitude response characteristics in dB and phase response characteristics is given. The shape and the width of pass-band of proposed novel classes of FIR filter function are studied in details for a given value of maximal attenuation in the filter pass-band. Also, the examples of stop-band of proposed ultra selective 2D FIR filter functions are illustrated. Parasitic Gibss effect is the main shortage of sloped interrupt of infinite Fourier order and for given examples it is evidently suppressed. The possible applications of these filters are for special telecommunication links as they have a very narrow pass-band and a good suppression in the stop-band. In addition, the proposed filters are of low complexity and with a high resolution and they are suitable for real-time applications. In addition, the proposed novel classes of 2D FIR filter functions can be very useful for future wireless applications and studies.

1. Introduction

Extremal property of classical orthonormal polynomials is shown by finite Christoffel-Darboux formula [1, 2]. The finite Christoffel-Darboux formula (summed from zero to n-th component) for the Chebyshev orthormal polynomials of the first kind has a form:

$$T_0(x)T_0(x) + T_1(x)T_1(x) + T_2(x)T_2(x) + \ldots + T_n(x)T_n(x) \tag{1}$$

For us, a normalized function $\Phi_1(n,x)$ for the Chebyshev orthormal polynomials of the first kind is of interest:

$$\Phi_1(n,x) = 1 + \frac{\sum_{r=0}^{n} T_r(x)T_r(x) - \sum_{r=0}^{n} T_r(0)T_r(0)}{\sum_{r=0}^{n} T_r(1)T_r(1) - \sum_{r=0}^{n} T_r(0)T_r(0)} \tag{2}$$

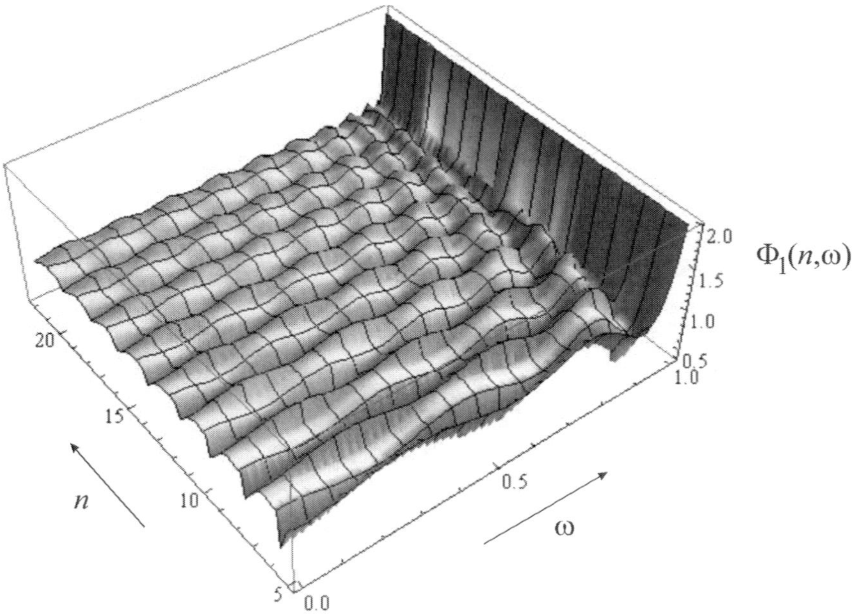

Figure 1. Normalized Christoffel-Darboux formula Φ1(n,x) for the Chebyshev orthormal polynomials of the first kind.

The finite Christoffel-Darboux formula (summed from zero to n-th component) for the Chebyshev orthormal polynomials of the second kind has a form:

$$U_0(x)U_0(x) + U_1(x)U_1(x) + U_2(x)U_2(x) + \ldots + U_n(x)U_n(x) \tag{3}$$

Normalized function $\Phi_2(n,x)$ for the Chebyshev orthormal polynomials of the second kind is of the form:

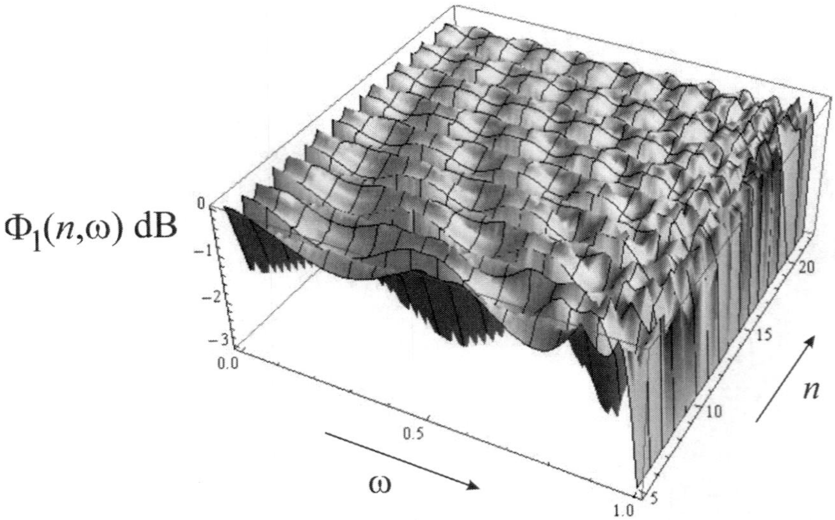

Figure 2. Figure 1 in dB.

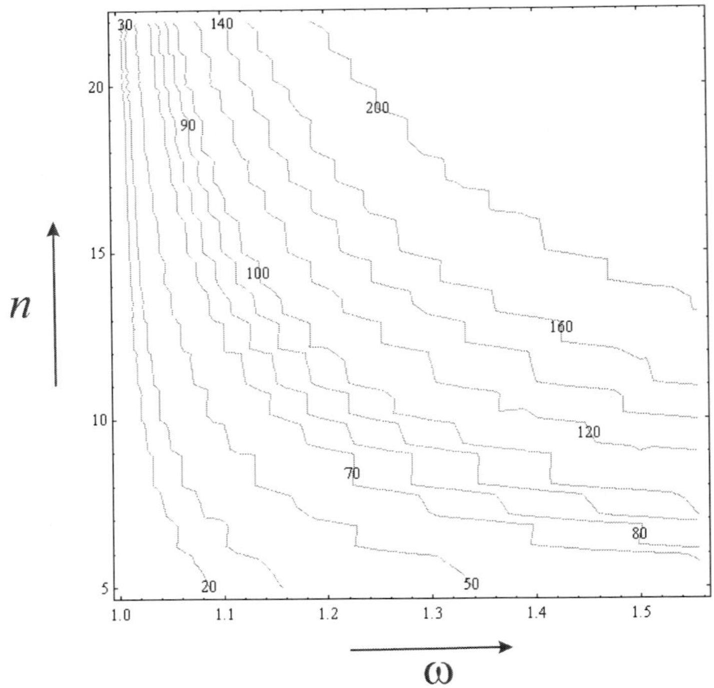

Figure 3. 2D contour plot of Figure 2 in dB.

$$\Phi_2(n,x) = 1 + \frac{\sum_{r=0}^{n} U_r(x)U_r(x) - \sum_{r=0}^{n} U_r(0)U_r(0)}{\sum_{r=0}^{n} U_r(1)U_r(1) - \sum_{r=0}^{n} U_r(0)U_r(0)} \tag{4}$$

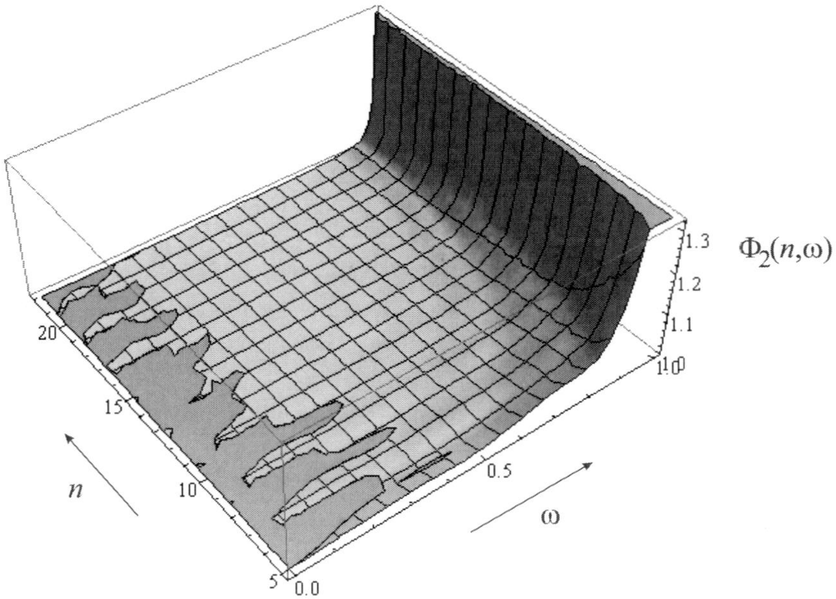

Figure 4. Normalized Christoffel-Darboux formula Φ2(n,x) for the Chebyshev orthormal polynomials of the second kind.

Continual extremal finite Christoffel-Darboux sum for classical orthormal polynomials has found an application in the theory of electrical filters. Continual LC filters either active or passive are obtained by most direct application of Christoffel-Darboux sum for the Lagandre polynomials, the Chebyshev polynomials of the first kind, the Chebyshev polynomials of the second kind, the Gegenbauer polynomials and the Jacobi polynomials. These filter functions are obtained in a compact explicit form by using analytical method [3-7]. Special significance has the Christoffel-Darboux formula for the Chebyshev polynomials of the first kind, shown in Figures 1, 2 and 3 (3D plot and 2D contour plot) and for the Chebyshev polynomials of the second kind, shown in Figures 4, 5 and 6 (3D plot and 2D contour plot). By correct mapping without errors and with residues, novel classes of filter functions in

1D z domain and in 2D z domain can be obtained in analytical form. 1D FIR filter functions in z domain of low complexity, with linear phase and multiplierless realization are given in literature [8-12]. All circuits which do not contain multipliers and with linear phase have a very distinguish advantage for correct research of unknown phenomena because the effect of finite length of filter coefficients due to finite quantization of coefficients is eliminated. Analytical methods for multiplierless linear phase 2D FIR filter functions are given in literature [12-14]. There are other methods for design of low-pass miltiplierless FIR filter functions such as using computer algebra system [15, 16]. Classical CIC filters have found numerous applications and gained a high popularity in modern wireless communications as shown in [17-21].

In this chapter, three novel classes of selective non-recursive 2D FIR filter functions are proposed for the special professional wireless applications. Examples of filter functions, coefficients of impulse response, amplitude, magnitude and linear phase response characteristics are given here. For considered examples, the width of stop-band in percentages is given in tables for the set of attenuation values in the stop-band. Also, the width of normalized pass-band is given for the attenuation value of 0.28 dB in the pass-band. Selective filters with a very narrow pass-band, whose location can be in 2D z-domain for the first two classes or at coordinate origin of 2D z-domain for the third class, are proposed.

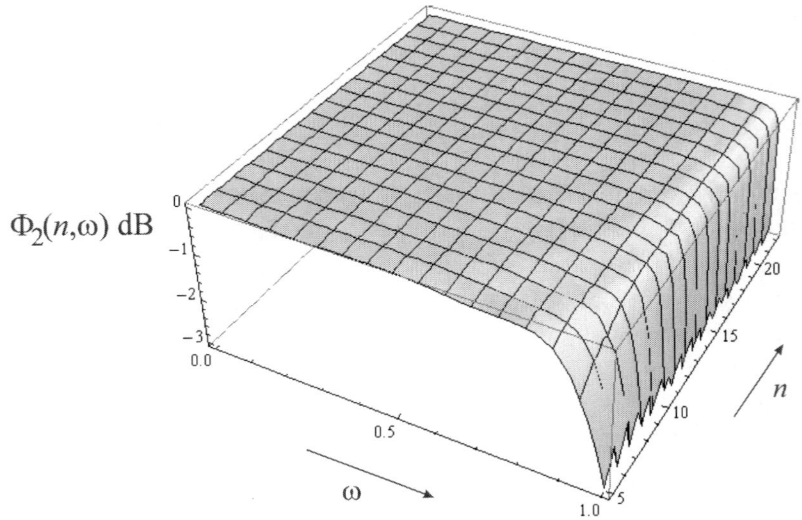

Figure 5. Figure 4 in dB.

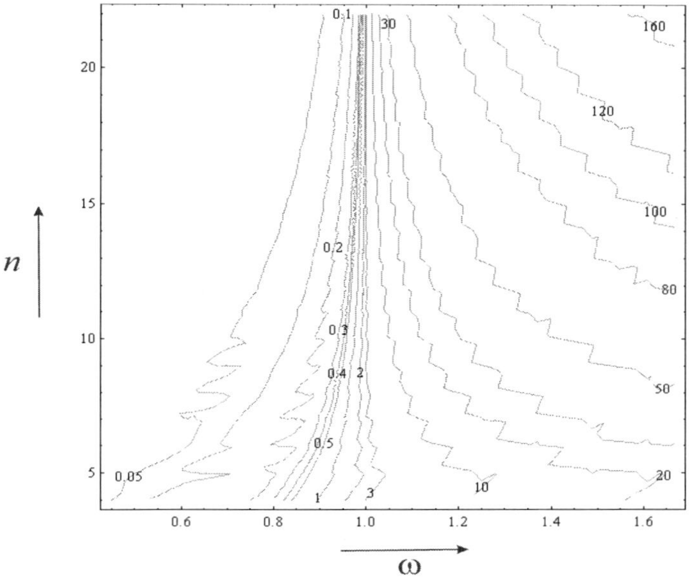

Figure 6. 2D contour plot of Figure 5 in dB.

2. 2D FIR FILTER FUNCTION

A linear phase 2D symmetric FIR filter of $N*N$-order is defined by:

$$H(N, z_1, z_2) = K_1 \sum_{r=0}^{N} \sum_{k=0}^{N} h(N,r,k)\, z_1^{-r}\, z_2^{-k} , \tag{5}$$

where K_1 is the gain real constant and $h(N,r,k)$ are the coefficients of impulse response that are real numbers. Square of the filter frequency response can be represented by:

$$H(N, z_1, z_2) H(N, \overline{z_1}, \overline{z_2}), \text{ for } z_1 \rightarrow e^{j\omega_1},\ z_2 \rightarrow e^{j\omega_2} \tag{6}$$

Or

$$\left| H\left(N, e^{j\omega_1}, e^{j\omega_2}\right) \right|^2 = \left| H\left(N, e^{j\omega_1}, e^{j\omega_2}\right) H\left(N, e^{-j\omega_1}, e^{-j\omega_2}\right) \right| \tag{7}$$

Alternatively in dB, squared filter frequency response can be presented by:

$$20\log\left|H\left(N,e^{j\omega_1},e^{j\omega_2}\right)\right|, \quad [\mathrm{dB}]. \tag{8}$$

3. The First Novel Class of Non-Recursive 2D FIR Filter Functions

In this chapter we propose the first novel class of non-causal linear phase and non-recursive 2D symmetric FIR filter functions with dimension $N*N$ in the following explicit form:

$$H_1\left(N,z_1,z_2\right)=K_1\sum_{r=1}^{N}(z_1^{+2r}-2+z_1^{-2r})\times(z_2^{+2r}+2+z_2^{-2r}) \tag{9}$$

The first novel class of causal 2D FIR filter function can be generated from Eq. (9) by its simple multiplication with factor $z_1^{-2N}*z_2^{-2N}$ as:

$$H_1(N,z_1,z_2)=K_1\sum_{r=1}^{N}(z_1^{-2N+2r}-2z_1^{-2N}+z_1^{-2N-2r})*(z_2^{-2N+2r}+2z_2^{-2N}+z_2^{-2N-2r}) \tag{10}$$

Frequency response characteristics, amplitude response characteristics, magnitude response characteristics and linear phase response characteristics of the first novel class of 2D symmetric FIR filter functions are given by the following equations:

$$H_1\left(N,e^{i\omega_1},e^{j\omega_2}\right)=K_1\,e^{-i2N\omega_1}e^{-j2N\omega_2}\sum_{r=1}^{N}(\cos(2r\omega_1)-1)\,(\cos(2r\omega_2)+1), \tag{11}$$

$$A_1\left(N,\omega_1,\omega_2\right)=\sum_{r=1}^{N}(\cos(2r\omega_1)-1)\,(\cos(2r\omega_2)+1), \tag{12}$$

$$\left|H_1\left(N,e^{i\omega_1},e^{j\omega_2}\right)\right|=\left|A_1(N,\omega_1,\omega_2)\right|, \tag{13}$$

$$\varphi_1(N,\omega_1,\omega_2) = -i2N\omega_1 - j2N\omega_2, \tag{14}$$

respectively.

The frequency response characteristics in the z domain for N=5 is:

$$\begin{aligned} &H_1(N=5,z_1,z_2) = 1 + z_1^{-20} + 2z_1^{-10} + z_2^{-20} + z_1^{-20}z_2^{-20} + 2z_1^{-10}z_2^{-20} + z_1^{-18}z_2^{-18} + \\ &2z_1^{-10}z_2^{-18} + z_1^{-2}z_2^{-18} + z_1^{-16}z_2^{-16} + 2z_1^{-10}z_2^{-16} + z_1^{-4}z_2^{-16} + z_1^{-14}z_2^{-14} + 2z_1^{-10}z_2^{-14} + \\ &z_1^{-6}z_2^{-14} + z_1^{-12}z_2^{-12} + 2z_1^{-10}z_2^{-12} + z_1^{-8}z_2^{-12} + 2z_2^{-10} - 2z_1^{-20}z_2^{-10} - 2z_1^{-18}z_2^{-10} - \\ &2z_1^{-16}z_2^{-10} - 2z_1^{-14}z_2^{-10} - 2z_1^{-12}z_2^{-10} - 20z_1^{-10}z_2^{-10} - 2z_1^{-8}z_2^{-10} - 2z_1^{-6}z_2^{-10} - 2z_1^{-4}z_2^{-10} - \\ &2z_1^{-2}z_2^{-10} + z_1^{-12}z_2^{-8} + 2z_1^{-10}z_2^{-8} + z_1^{-8}z_2^{-8} + z_1^{-14}z_2^{-6} + 2z_1^{-10}z_2^{-6} + z_1^{-6}z_2^{-6} + \\ &z_1^{-16}z_2^{-4} + 2z_1^{-10}z_2^{-4} + z_1^{-4}z_2^{-4} + z_1^{-18}z_2^{-2} + 2z_1^{-10}z_2^{-2} + z_1^{-2}z_2^{-2} \end{aligned} \tag{15}$$

The coefficients of impulse response of the first novel class of the 2D FIR filter functions and their 2D contour plot, for N=5, are given in Figure 7 and Figure 8, respectively.

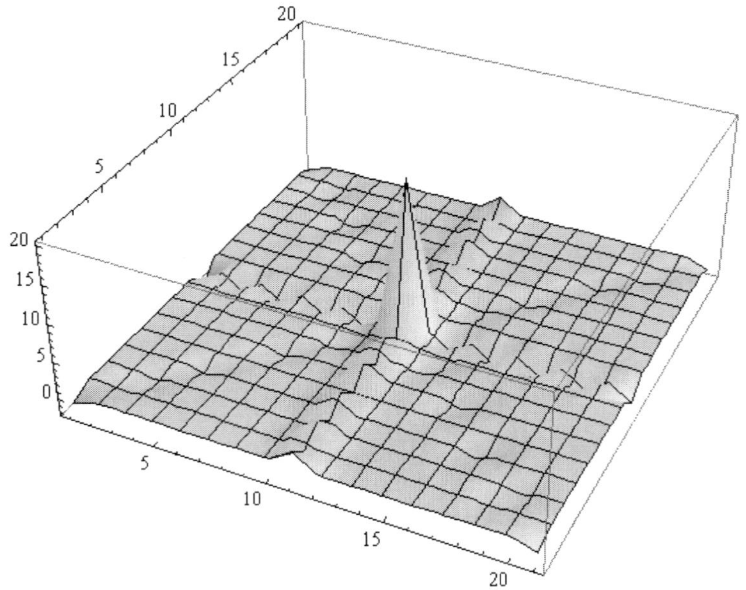

Figure 7. Coefficients of impulse response h1(N=5,r,k) of the first novel class of the 2D FIR filter functions.

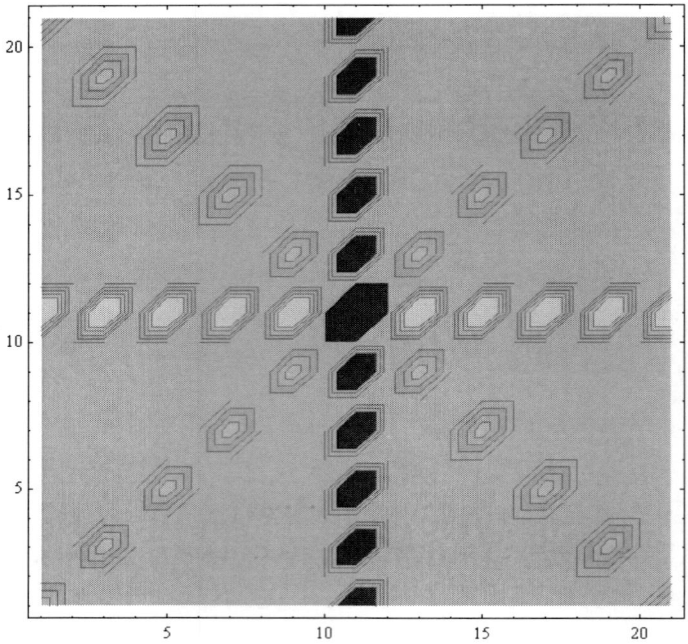

Figure 8. 2D contour plot of coefficients of impulse response $h_1(N=5,r,k)$ of the first novel class of the 2D FIR filter function.

The values of coefficients of impulse response $h_1(N=5,r,k)$ of the first novel class of the 2D FIR filter functions are given below:

```
h1(N=5,r,k)=
{{1,0,0,0,0,0,0,0,0,-2,0,0,0,0,0,0,0,0,1},
{0,0,0,0,0,0,0,0,0,0,0,0,0,0,0,0,0,0,0},
{0,0,1,0,0,0,0,0,0,0,-2,0,0,0,0,0,0,1,0,0},
{0,0,0,0,0,0,0,0,0,0,0,0,0,0,0,0,0,0,0},
{0,0,0,0,1,0,0,0,0,0,-2,0,0,0,0,0,1,0,0,0,0},
{0,0,0,0,0,0,0,0,0,0,0,0,0,0,0,0,0,0,0},
{0,0,0,0,0,0,1,0,0,0,-2,0,0,0,1,0,0,0,0,0,0},
{0,0,0,0,0,0,0,0,0,0,0,0,0,0,0,0,0,0,0},
{0,0,0,0,0,0,0,0,1,0,-2,0,1,0,0,0,0,0,0,0,0},
{0,0,0,0,0,0,0,0,0,0,0,0,0,0,0,0,0,0,0},
{2,0,2,0,2,0,2,0,2,0,-20,0,2,0,2,0,2,0,2,0,2},
{0,0,0,0,0,0,0,0,0,0,0,0,0,0,0,0,0,0,0},
{0,0,0,0,0,0,0,0,1,0,-2,0,1,0,0,0,0,0,0,0,0},
```

{0,0,0,0,0,0,0,0,0,0,0,0,0,0,0,0,0,0,0,0,0,0,0,0,0},
{0,0,0,0,0,0,1,0,0,0,0,-2,0,0,0,1,0,0,0,0,0,0},
{0,0,0,0,0,0,0,0,0,0,0,0,0,0,0,0,0,0,0,0,0,0,0,0,0},
{0,0,0,0,1,0,0,0,0,0,0,-2,0,0,0,0,0,1,0,0,0,0,0},
{0,0,0,0,0,0,0,0,0,0,0,0,0,0,0,0,0,0,0,0,0,0,0,0,0},
{0,0,1,0,0,0,0,0,0,0,0,-2,0,0,0,0,0,0,0,1,0,0},
{0,0,0,0,0,0,0,0,0,0,0,0,0,0,0,0,0,0,0,0,0,0,0,0,0},
{1,0,0,0,0,0,0,0,0,0,0,-2,0,0,0,0,0,0,0,0,0,0,1}}. (16)

The frequency response characteristics in the z domain for N=6 is:

$$\begin{aligned}
&H_1(N=6,z_1,z_2)=1+z_1^{-24}-2z_1^{-12}+z_2^{-24}+z_1^{-24}z_2^{-24}-2z_1^{-12}z_2^{-24}+z_1^{-22}z_2^{-22}-\\
&2z_1^{-12}z_2^{-22}+z_1^{-2}z_2^{-22}+z_1^{-20}z_2^{-20}-2z_1^{-12}z_2^{-20}+z_1^{-4}z_2^{-20}+z_1^{-18}z_2^{-18}-2z_1^{-12}z_2^{-18}+\\
&z_1^{-6}z_2^{-18}+z_1^{-16}z_2^{-16}-2z_1^{-12}z_2^{-16}+z_1^{-8}z_2^{-16}+z_1^{-14}z_2^{-14}-2z_1^{-12}z_2^{-14}+z_1^{-10}z_2^{-14}+\\
&2z_1^{-12}+2z_1^{-24}z_2^{-12}+2z_1^{-22}z_2^{-12}+2z_1^{-20}z_2^{-12}+2z_1^{-18}z_2^{-12}+2z_1^{-16}z_2^{-12}+2z_1^{-14}z_2^{-12}-\\
&24z_1^{-12}z_2^{-12}+2z_1^{-10}z_2^{-12}+2z_1^{-8}z_2^{-12}+2z_1^{-6}z_2^{-12}+2z_1^{-4}z_2^{-12}+2z_1^{-2}z_2^{-12}+z_1^{-14}z_2^{-10}-\\
&2z_1^{-12}z_2^{-10}+z_1^{-10}z_2^{-10}+z_1^{-16}z_2^{-8}-2z_1^{-12}z_2^{-8}+z_1^{-8}z_2^{-8}+z_1^{-18}z_2^{-6}-2z_1^{-12}z_2^{-6}+\\
&z_1^{-6}z_2^{-6}+z_1^{-20}z_2^{-4}-2z_1^{-12}z_2^{-4}+z_1^{-4}z_2^{-4}+z_1^{-22}z_2^{-2}-2z_1^{-12}z_2^{-2}+z_1^{-2}z_2^{-2}
\end{aligned}$$

(17)

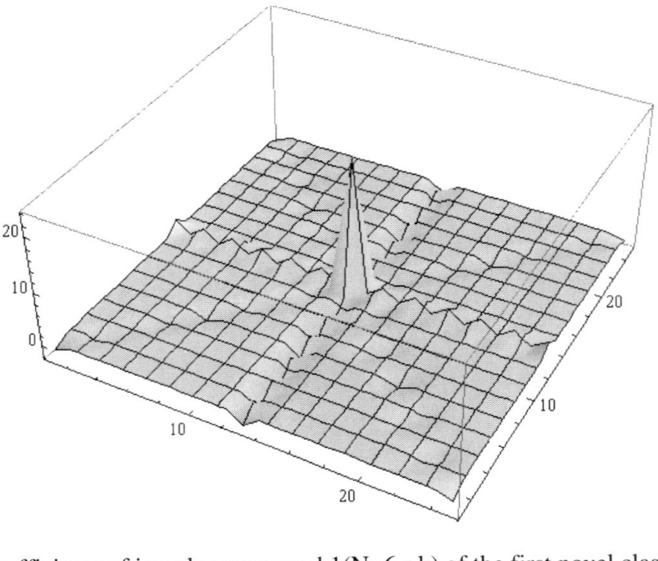

Figure 9. Coefficients of impulse response h1(N=6,r,k) of the first novel class of the 2D FIR filter functions.

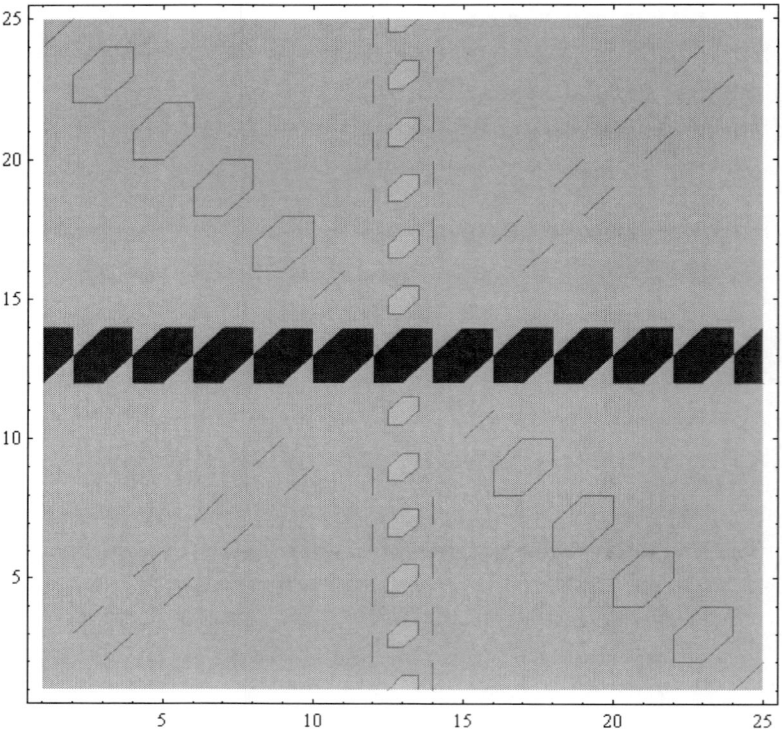

Figure 10. 2D contour plot of coefficients of impulse response h1(N=6,r,k) of the first novel class of the 2D FIR filter functions.

The coefficients of impulse response of the first novel class of the 2D FIR filter functions and their contour plot, for N=6, are given in Figure 9 and Figure 10, respectively.

The values of coefficients of impulse response h1(N=6,r,k) of the first novel class of the 2D FIR filter functions are given below:

h1(N=6,r,k)=
{{1,0,0,0,0,0,0,0,0,0,0,0,2,0,0,0,0,0,0,0,0,0,0,0,1},
{0,0,0,0,0,0,0,0,0,0,0,0,0,0,0,0,0,0,0,0,0,0,0,0,0},
{0,0,1,0,0,0,0,0,0,0,0,0,2,0,0,0,0,0,0,0,0,0,1,0,0},
{0,0,0,0,0,0,0,0,0,0,0,0,0,0,0,0,0,0,0,0,0,0,0,0,0},
{0,0,0,0,1,0,0,0,0,0,0,0,2,0,0,0,0,0,0,0,1,0,0,0,0},
{0,0,0,0,0,0,0,0,0,0,0,0,0,0,0,0,0,0,0,0,0,0,0,0,0},
{0,0,0,0,0,0,1,0,0,0,0,0,2,0,0,0,0,0,1,0,0,0,0,0,0},
{0,0,0,0,0,0,0,0,0,0,0,0,0,0,0,0,0,0,0,0,0,0,0,0,0},

{0,0,0,0,0,0,0,1,0,0,0,2,0,0,0,1,0,0,0,0,0,0,0},
{0,0,0,0,0,0,0,0,0,0,0,0,0,0,0,0,0,0,0,0,0,0,0},
{0,0,0,0,0,0,0,0,0,1,0,2,0,1,0,0,0,0,0,0,0,0,0},
{0,0,0,0,0,0,0,0,0,0,0,0,0,0,0,0,0,0,0,0,0,0,0},
{-2,0,-2,0,-2,0,-2,0,-2,0,-2,0,-24,0,-2,0,-2,0,-2,0,-2,0,-2,0,-2},
{0,0,0,0,0,0,0,0,0,0,0,0,0,0,0,0,0,0,0,0,0,0,0},
{0,0,0,0,0,0,0,0,0,1,0,2,0,1,0,0,0,0,0,0,0,0,0},
{0,0,0,0,0,0,0,0,0,0,0,0,0,0,0,0,0,0,0,0,0,0,0},
{0,0,0,0,0,0,0,1,0,0,0,2,0,0,0,1,0,0,0,0,0,0,0},
{0,0,0,0,0,0,0,0,0,0,0,0,0,0,0,0,0,0,0,0,0,0,0},
{0,0,0,0,0,0,1,0,0,0,0,0,2,0,0,0,0,0,1,0,0,0,0,0,0},
{0,0,0,0,0,0,0,0,0,0,0,0,0,0,0,0,0,0,0,0,0,0,0},
{0,0,0,0,1,0,0,0,0,0,0,0,2,0,0,0,0,0,0,0,1,0,0,0,0},
{0,0,0,0,0,0,0,0,0,0,0,0,0,0,0,0,0,0,0,0,0,0,0},
{0,0,1,0,0,0,0,0,0,0,0,0,2,0,0,0,0,0,0,0,0,0,1,0,0},
{0,0,0,0,0,0,0,0,0,0,0,0,0,0,0,0,0,0,0,0,0,0,0},
{1,0,0,0,0,0,0,0,0,0,0,0,2,0,0,0,0,0,0,0,0,0,0,0,1}}. (18)

The frequency response characteristics in the z domain for N=9 is:

$$
\begin{aligned}
&H_1(N=9,z_1,z_2)=1+z_1^{-36}+2z_1^{-18}+z_2^{-36}+z_1^{-36}z_2^{-36}+2z_1^{-18}z_2^{-36}+z_1^{-34}z_2^{-34}+\\
&2z_1^{-18}z_2^{-34}+z_1^{-2}z_2^{-34}+z_1^{-32}z_2^{-32}+2z_1^{-18}z_2^{-32}+z_1^{-4}z_2^{-32}+z_1^{-30}z_2^{-30}+2z_1^{-18}z_2^{-30}+\\
&z_1^{-6}z_2^{-30}+z_1^{-28}z_2^{-28}+2z_1^{-18}z_2^{-28}+z_1^{-8}z_2^{-28}+z_1^{-26}z_2^{-26}+2z_1^{-18}z_2^{-26}+z_1^{-10}z_2^{-26}+\\
&z_1^{-24}z_2^{-24}+2z_1^{-18}z_2^{-24}+z_1^{-12}z_2^{-24}+z_1^{-22}z_2^{-22}+2z_1^{-18}z_2^{-22}+z_1^{-14}z_2^{-22}+z_1^{-20}z_2^{-20}+\\
&2z_1^{-18}z_2^{-20}+z_1^{-16}z_2^{-20}-2z_2^{-18}-2z_1^{-36}z_2^{-18}-2z_1^{-34}z_2^{-18}-2z_1^{-32}z_2^{-18}-2z_1^{-30}z_2^{-18}-\\
&2z_1^{-28}z_2^{-18}-2z_1^{-26}z_2^{-18}-2z_1^{-24}z_2^{-18}-2z_1^{-22}z_2^{-18}-2z_1^{-20}z_2^{-18}-36z_1^{-18}z_2^{-18}-2z_1^{-16}z_2^{-18}-\\
&2z_1^{-14}z_2^{-18}-2z_1^{-12}z_2^{-18}-2z_1^{-10}z_2^{-18}-2z_1^{-8}z_2^{-18}-2z_1^{-6}z_2^{-18}-2z_1^{-4}z_2^{-18}-2z_1^{-2}z_2^{-18}+\\
&z_1^{-20}z_2^{-16}+2z_1^{-18}z_2^{-16}+z_1^{-16}z_2^{-16}+z_1^{-22}z_2^{-14}+2z_1^{-18}z_2^{-14}+z_1^{-14}z_2^{-14}+z_1^{-24}z_2^{-14}+\\
&2z_1^{-18}z_2^{-12}+z_1^{-12}z_2^{-12}+z_1^{-26}z_2^{-10}+2z_1^{-18}z_2^{-10}+z_1^{-10}z_2^{-10}+z_1^{-28}z_2^{-8}+2z_1^{-18}z_2^{-8}+\\
&z_1^{-8}z_2^{-8}+z_1^{-30}z_2^{-6}+2z_1^{-18}z_2^{-6}+z_1^{-6}z_2^{-6}+z_1^{-32}z_2^{-4}+2z_1^{-18}z_2^{-4}+z_1^{-4}z_2^{-4}+\\
&z_1^{-34}z_2^{-2}+2z_1^{-18}z_2^{-2}+z_1^{-2}z_2^{-2}
\end{aligned}
$$
(19)

The coefficients of impulse response of the first novel class of the 2D FIR filter functions and their contour plot, for N=9, are given in Figure 11 and Figure 12, respectively.

The values of coefficients of impulse response h1(N=9,r,k) of the first novel class of the 2D FIR filter functions are given below:

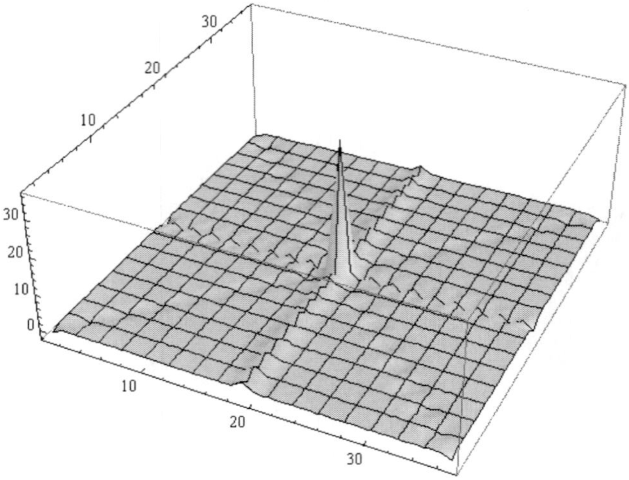

Figure 11. Coefficients of impulse response h1(N=9,r,k) of the first novel class of the 2D FIR filter functions.

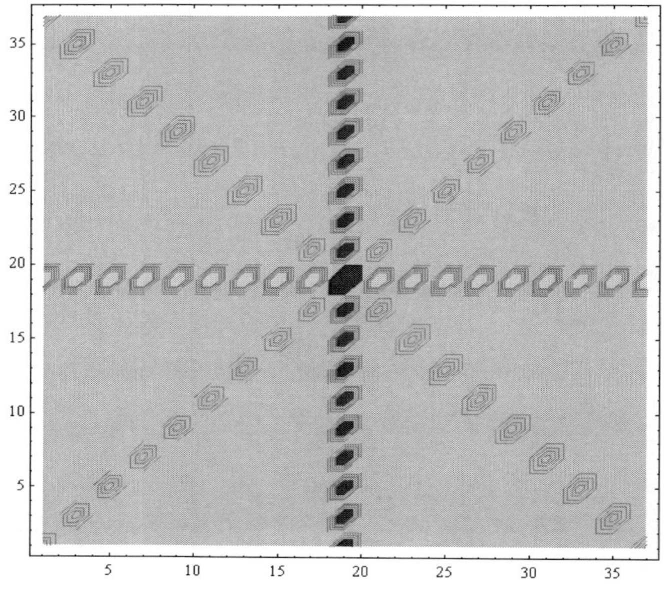

Figure 12. 2D contour plot of coefficients of impulse response h1(N=9,r,k) of the first novel class of the 2D FIR filter functions.

$h_1(N=9,r,k)=$

{{1,0,0,0,0,0,0,0,0,0,0,0,0,0,0,0,0,0,-2,0,0,0,0,0,0,0,0,0,0,0,0,0,0,0,0,0,1},
{0,0,0,0,0,0,0,0,0,0,0,0,0,0,0,0,0,0,0,0,0,0,0,0,0,0,0,0,0,0,0,0,0,0,0,0,0},
{0,0,1,0,0,0,0,0,0,0,0,0,0,0,0,0,0,0,-2,0,0,0,0,0,0,0,0,0,0,0,0,0,0,0,1,0,0},
{0,0,0,0,0,0,0,0,0,0,0,0,0,0,0,0,0,0,0,0,0,0,0,0,0,0,0,0,0,0,0,0,0,0,0,0,0},
{0,0,0,0,1,0,0,0,0,0,0,0,0,0,0,0,0,0,-2,0,0,0,0,0,0,0,0,0,0,0,0,0,1,0,0,0,0},
{0,0,0,0,0,0,0,0,0,0,0,0,0,0,0,0,0,0,0,0,0,0,0,0,0,0,0,0,0,0,0,0,0,0,0,0,0},
{0,0,0,0,0,0,1,0,0,0,0,0,0,0,0,0,0,0,-2,0,0,0,0,0,0,0,0,0,0,0,1,0,0,0,0,0,0},
{0,0,0,0,0,0,0,0,0,0,0,0,0,0,0,0,0,0,0,0,0,0,0,0,0,0,0,0,0,0,0,0,0,0,0,0,0},
{0,0,0,0,0,0,0,0,1,0,0,0,0,0,0,0,0,0,-2,0,0,0,0,0,0,0,0,0,1,0,0,0,0,0,0,0,0},
{0,0,0,0,0,0,0,0,0,0,0,0,0,0,0,0,0,0,0,0,0,0,0,0,0,0,0,0,0,0,0,0,0,0,0,0,0},
{0,0,0,0,0,0,0,0,0,0,1,0,0,0,0,0,0,0,-2,0,0,0,0,0,0,0,1,0,0,0,0,0,0,0,0,0,0},
{0,0,0,0,0,0,0,0,0,0,0,0,0,0,0,0,0,0,0,0,0,0,0,0,0,0,0,0,0,0,0,0,0,0,0,0,0},
{0,0,0,0,0,0,0,0,0,0,0,0,1,0,0,0,0,0,-2,0,0,0,0,0,1,0,0,0,0,0,0,0,0,0,0,0,0},
{0,0,0,0,0,0,0,0,0,0,0,0,0,0,0,0,0,0,0,0,0,0,0,0,0,0,0,0,0,0,0,0,0,0,0,0,0},
{0,0,0,0,0,0,0,0,0,0,0,0,0,0,1,0,0,0,-2,0,0,0,1,0,0,0,0,0,0,0,0,0,0,0,0,0,0},
{0,0,0,0,0,0,0,0,0,0,0,0,0,0,0,0,0,0,0,0,0,0,0,0,0,0,0,0,0,0,0,0,0,0,0,0,0},
{0,0,0,0,0,0,0,0,0,0,0,0,0,0,0,0,1,0,-2,0,1,0,0,0,0,0,0,0,0,0,0,0,0,0,0,0,0},
{0,0,0,0,0,0,0,0,0,0,0,0,0,0,0,0,0,0,0,0,0,0,0,0,0,0,0,0,0,0,0,0,0,0,0,0,0},
{2,0,2,0,2,0,2,0,2,0,2,0,2,0,2,0,2,0,-36,0,2,0,2,0,2,0,2,0,2,0,2,0,2,0,2,0,2},
{0,0,0,0,0,0,0,0,0,0,0,0,0,0,0,0,0,0,0,0,0,0,0,0,0,0,0,0,0,0,0,0,0,0,0,0,0},
{0,0,0,0,0,0,0,0,0,0,0,0,0,0,0,0,1,0,-2,0,1,0,0,0,0,0,0,0,0,0,0,0,0,0,0,0,0},
{0,0,0,0,0,0,0,0,0,0,0,0,0,0,0,0,0,0,0,0,0,0,0,0,0,0,0,0,0,0,0,0,0,0,0,0,0},
{0,0,0,0,0,0,0,0,0,0,0,0,0,0,1,0,0,0,-2,0,0,0,1,0,0,0,0,0,0,0,0,0,0,0,0,0,0},
{0,0,0,0,0,0,0,0,0,0,0,0,0,0,0,0,0,0,0,0,0,0,0,0,0,0,0,0,0,0,0,0,0,0,0,0,0},
{0,0,0,0,0,0,0,0,0,0,0,0,1,0,0,0,0,0,-2,0,0,0,0,0,1,0,0,0,0,0,0,0,0,0,0,0,0},
{0,0,0,0,0,0,0,0,0,0,0,0,0,0,0,0,0,0,0,0,0,0,0,0,0,0,0,0,0,0,0,0,0,0,0,0,0},
{0,0,0,0,0,0,0,0,0,0,1,0,0,0,0,0,0,0,-2,0,0,0,0,0,0,0,1,0,0,0,0,0,0,0,0,0,0},
{0,0,0,0,0,0,0,0,0,0,0,0,0,0,0,0,0,0,0,0,0,0,0,0,0,0,0,0,0,0,0,0,0,0,0,0,0},
{0,0,0,0,0,0,0,0,1,0,0,0,0,0,0,0,0,0,-2,0,0,0,0,0,0,0,0,0,1,0,0,0,0,0,0,0,0},
{0,0,0,0,0,0,0,0,0,0,0,0,0,0,0,0,0,0,0,0,0,0,0,0,0,0,0,0,0,0,0,0,0,0,0,0,0},
{0,0,0,0,0,0,1,0,0,0,0,0,0,0,0,0,0,0,-2,0,0,0,0,0,0,0,0,0,0,0,1,0,0,0,0,0,0},
{0,0,0,0,0,0,0,0,0,0,0,0,0,0,0,0,0,0,0,0,0,0,0,0,0,0,0,0,0,0,0,0,0,0,0,0,0},
{0,0,0,0,1,0,0,0,0,0,0,0,0,0,0,0,0,0,-2,0,0,0,0,0,0,0,0,0,0,0,0,0,1,0,0,0,0},
{0,0,0,0,0,0,0,0,0,0,0,0,0,0,0,0,0,0,0,0,0,0,0,0,0,0,0,0,0,0,0,0,0,0,0,0,0},
{0,0,1,0,0,0,0,0,0,0,0,0,0,0,0,0,0,0,-2,0,0,0,0,0,0,0,0,0,0,0,0,0,0,0,1,0,0},
{0,0,0,0,0,0,0,0,0,0,0,0,0,0,0,0,0,0,0,0,0,0,0,0,0,0,0,0,0,0,0,0,0,0,0,0,0},
{1,0,0,0,0,0,0,0,0,0,0,0,0,0,0,0,0,0,-2,0,0,0,0,0,0,0,0,0,0,0,0,0,0,0,0,0,1}}

(20)

For the linear phase 2D symmetric FIR digital filters generated by the proposed technique, the following symmetries are valid:

$$H_1\left(N,z_1,-z_2\right)=H_1\left(N,z_1,z_2\right), \tag{21}$$

$$H_1\left(N,-z_1,z_2\right)=H_1\left(N,z_1,z_2\right). \tag{22}$$

The linear phase function of the 2D symmetric FIR digital filter defined by Eq. (11) has the following form for $z_1 \to e^{i\omega_1}$ and $z_2 \to e^{j\omega_2}$:

$$\varphi_1(N,\omega_1,\omega_2) = -i2N\omega_1 - j2N\omega_2 . \tag{23}$$

Similarly to the approach used in the literature [12], a filter function of K cascaded blocks $H\left(N,z_1,z_2\right)$, given by Eq. (10), for an efficient suppression of unwanted parasitic Gibss phenomena, of the form:

$$\left(H\left(N,z_1,z_2\right)\right)^K, \tag{24}$$

representing multiple zero function is considered here. In addition, we propose that the function $H\left(N,z_1,z_2\right)$ is realized as a product of three functions of successive orders $M-1$, M and $M+1$:

$$H_1\left(M-1,z_1,z_2\right)H_1\left(M,z_1,z_2\right)H_1\left(M+1,z_1,z_2\right). \tag{25}$$

So, general form of the proposed filter function is given by the following expression:

$$H_1\left(M,K,z_1,z_2\right)=\left[H_1\left(M-1,z_1,z_2\right)H_1\left(M,z_1,z_2\right)H_1\left(M+1,z_1,z_2\right)\right]^K \tag{26}$$

Frequency response characteristics of the first novel class of the 2D FIR filter functions, $H_1(M,K,e^{i\omega_1},e^{j\omega_2})$, obtained after inserting Eq.(10) in Eq.(26), is:

$$H_1(M,K,e^{i\omega_1},e^{j\omega_2}) = K_1\, e^{-i6MK\omega_1} e^{-j6MK\omega_2}$$
$$[\sum_{k=1}^{M-1}(\cos(2k\omega_1)-1)\,(\cos(2k\omega_2)+1)\sum_{r=1}^{M}(\cos(2r\omega_1)-1)\,(\cos(2r\omega_2)+1)$$
$$\sum_{m=1}^{M+1}(\cos(2m\omega_1)-1)\,(\cos(2m\omega_2)+1)]^K \tag{27}$$

Amplitude response characteristics, magnitude response characteristics and linear phase response characteristics of the first novel class of the 2D FIR filter function, obtained from Eq.(27), are of the following form:

$$A_1\left(M,K,\omega_1,\omega_2\right)=[\sum_{k=1}^{M-1}(\cos(2k\omega_1)-1)\,(\cos(2k\omega_2)+1)$$
$$\sum_{r=1}^{M}(\cos(2r\omega_1)-1)\,(\cos(2r\omega_2)+1)\sum_{m=1}^{M+1}(\cos(2m\omega_1)-1)\,(\cos(2m\omega_2)+1)]^K \tag{28}$$

$$\left|H_1\left(M,K,e^{j\omega_1},e^{j\omega_2}\right)\right| = \left|A_1(M,K,\omega_1,\omega_2)\right|, \tag{29}$$

$$\varphi_1(M,K,\omega_1,\omega_2) = -\,i6MK\omega_1 - j6MK\omega_2, \tag{30}$$

respectively.

4. The Second Novel Class of Non-Recursive 2D FIR Filter Functions

We propose the second novel class of non-causal 2D symmetric linear phase and non-recursive FIR filter functions with dimension $N*N$ in the following explicit form:

$$H_2\left(N, z_1, z_2\right) = K_1 \sum_{r=1}^{N} (z_1^{+2r} - 2 + z_1^{-2r}) \times (z_2^{+2r} + z_2^{-2r}) \tag{31}$$

The second novel class of causal 2D FIR filter function generated from Eq. (31) by simple multiplication with factor $z_1^{-2N} * z_2^{-2N}$ can be obtained as:

$$H_2(N, z_1, z_2) = K_1 \sum_{r=1}^{N} (z_1^{-2N+2r} - 2z_1^{-2N} + z_1^{-2N-2r}) \times (z_2^{-2N+2r} + z_2^{-2N-2r}) \tag{32}$$

Frequency response characteristics, amplitude response characteristics, magnitude response characteristics and linear phase response characteristics of the second novel class of 2D symmetric FIR filter functions are given by the following equations:

$$H_2(N, e^{j\omega_1}, e^{j\omega_2}) = K_1\, e^{i2N\omega_1} e^{j2N\omega_2} \sum_{r=1}^{N} (\cos(2r\omega_1) - 1)\ \cos(2r\omega_2) \tag{33}$$

$$A_2\left(N, \omega_1, \omega_2\right) = \sum_{r=1}^{N} (\cos(2r\omega_1) - 1)\ \cos(2r\omega_2) \tag{34}$$

$$\left| H_2\left(N, e^{i\omega_1}, e^{j\omega_2}\right) \right| = \left| A_2(N, \omega_1, \omega_2) \right|, \tag{35}$$

$$\varphi_2(N, \omega_1, \omega_2) = -i2N\omega_1 - j2N\omega_2, \tag{36}$$

respectively.

The frequency response characteristics in the z domain for N=5 is:

$$H_2(N=5,z_1,z_2)=1+z_1^{-20}-2z_1^{-10}+z_2^{-20}+z_1^{-20}z_2^{-20}-2z_1^{-10}z_2^{-20}+z_1^{-18}z_2^{-18}-2z_1^{-10}z_2^{-18}+z_1^{-2}z_2^{-18}+z_1^{-16}z_2^{-16}-2z_1^{-10}z_2^{-16}+z_1^{-4}z_2^{-16}+z_1^{-14}z_2^{-14}-2z_1^{-10}z_2^{-14}+z_1^{-6}z_2^{-14}+z_1^{-12}z_2^{-12}-2z_1^{-10}z_2^{-12}+z_1^{-8}z_2^{-12}+z_1^{-12}z_2^{-8}-2z_1^{-10}z_2^{-8}+z_1^{-8}z_2^{-8}+z_1^{-14}z_2^{-6}-2z_1^{-10}z_2^{-6}+z_1^{-6}z_2^{-6}+z_1^{-16}z_2^{-4}-2z_1^{-10}z_2^{-4}+z_1^{-4}z_2^{-4}+z_1^{-18}z_2^{-2}-2z_1^{-10}z_2^{-2}+z_1^{-2}z_2^{-2} \quad (37)$$

The coefficients of impulse response of the second novel class of the 2D FIR filter functions and their 2D contour plot, for N=5, are given in Figure 13 and Figure 14, respectively.

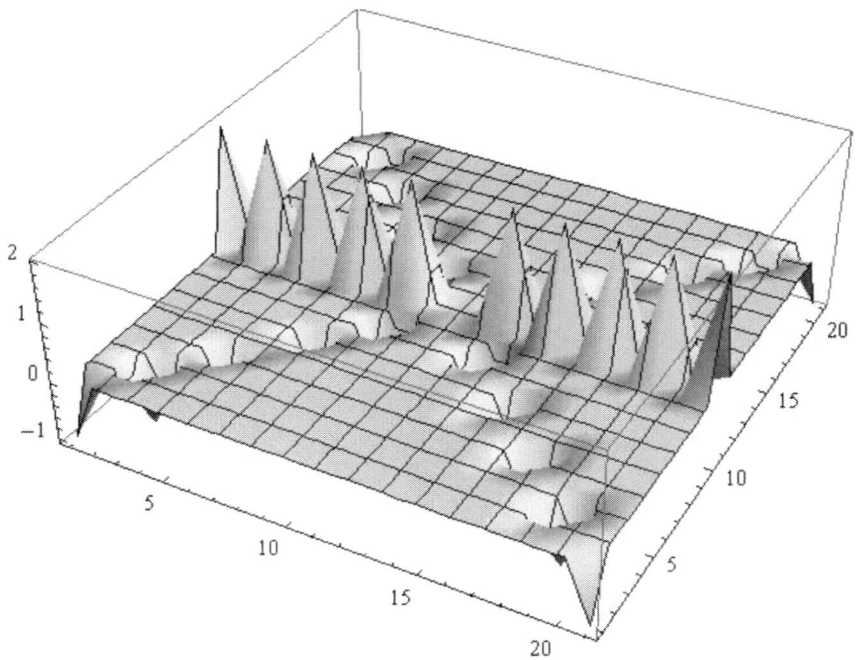

Figure 13. Coefficients of impulse response h2(N=5,r,k) of the second novel class of the 2D FIR filter functions.

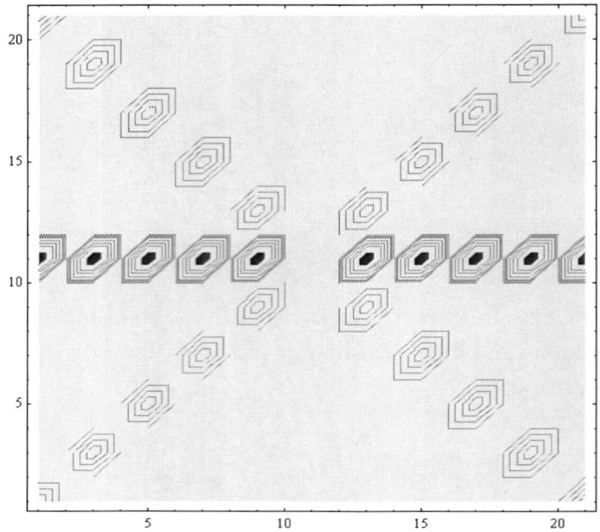

Figure 14. 2D contour plot of coefficients of impulse response h2(N=5,r,k) of the second novel class of the 2D FIR filter functions.

The values of coefficients of impulse response h2(N=5,r,k) of the second novel class of the 2D FIR filter function are given below:

h2(N=5,r,k)=
{{1,0,0,0,0,0,0,0,0,0,0,0,0,0,0,0,0,0,0,0,1},
{0,0,0,0,0,0,0,0,0,0,0,0,0,0,0,0,0,0,0,0,0},
{0,0,1,0,0,0,0,0,0,0,0,0,0,0,0,0,0,0,1,0,0},
{0,0,0,0,0,0,0,0,0,0,0,0,0,0,0,0,0,0,0,0,0},
{0,0,0,0,1,0,0,0,0,0,0,0,0,0,0,0,1,0,0,0,0},
{0,0,0,0,0,0,0,0,0,0,0,0,0,0,0,0,0,0,0,0,0},
{0,0,0,0,0,0,1,0,0,0,0,0,0,0,1,0,0,0,0,0,0},
{0,0,0,0,0,0,0,0,0,0,0,0,0,0,0,0,0,0,0,0,0},
{0,0,0,0,0,0,0,0,1,0,0,0,1,0,0,0,0,0,0,0,0},
{0,0,0,0,0,0,0,0,0,0,0,0,0,0,0,0,0,0,0,0,0},
{-2,0,-2,0,-2,0,-2,0,-2,0,0,0,-2,0,-2,0,-2,0,-2,0,-2},
{0,0,0,0,0,0,0,0,0,0,0,0,0,0,0,0,0,0,0,0,0},
{0,0,0,0,0,0,0,0,1,0,0,0,1,0,0,0,0,0,0,0,0},
{0,0,0,0,0,0,0,0,0,0,0,0,0,0,0,0,0,0,0,0,0},
{0,0,0,0,0,0,1,0,0,0,0,0,0,0,1,0,0,0,0,0,0},
{0,0,0,0,0,0,0,0,0,0,0,0,0,0,0,0,0,0,0,0,0},
{0,0,0,0,1,0,0,0,0,0,0,0,0,0,0,0,1,0,0,0,0},

{0,0,0,0,0,0,0,0,0,0,0,0,0,0,0,0,0,0,0,0,0,0,0,0,0},
{0,0,1,0,0,0,0,0,0,0,0,0,0,0,0,0,0,0,0,0,0,0,1,0,0},
{0,0,0,0,0,0,0,0,0,0,0,0,0,0,0,0,0,0,0,0,0,0,0,0,0},
{1,0,0,0,0,0,0,0,0,0,0,0,0,0,0,0,0,0,0,0,0,0,0,0,1}}. (38)

The frequency response characteristics in the z domain for N=6 is:

$$\begin{aligned}H_2(N=6,z_1,z_2) = {} & 1 + z_1^{-24} - 2z_1^{-12} + z_2^{-24} + z_1^{-24}z_2^{-24} - 2z_1^{-12}z_2^{-24} + z_1^{-22}z_2^{-22} - \\ & 2z_1^{-12}z_2^{-22} + z_1^{-2}z_2^{-22} + z_1^{-20}z_2^{-20} - 2z_1^{-12}z_2^{-20} + z_1^{-4}z_2^{-20} + z_1^{-18}z_2^{-18} - 2z_1^{-12}z_2^{-18} + \\ & z_1^{-6}z_2^{-18} + z_1^{-16}z_2^{-16} - 2z_1^{-12}z_2^{-16} + z_1^{-8}z_2^{-16} + z_1^{-14}z_2^{-14} - 2z_1^{-12}z_2^{-14} + z_1^{-10}z_2^{-14} + \\ & z_1^{-14}z_2^{-10} - 2z_1^{-12}z_2^{-10} + z_1^{-10}z_2^{-10} + z_1^{-16}z_2^{-8} - 2z_1^{-12}z_2^{-8} + z_1^{-8}z_2^{-8} + z_1^{-18}z_2^{-6} - \\ & 2z_1^{-12}z_2^{-6} + z_1^{-6}z_2^{-6} + z_1^{-20}z_2^{-4} - 2z_1^{-12}z_2^{-4} + z_1^{-4}z_2^{-4} + z_1^{-22}z_2^{-2} - 2z_1^{-12}z_2^{-2} + z_1^{-2}z_2^{-2}\end{aligned} \tag{39}$$

The coefficients of impulse response of the second novel class of the 2D FIR filter functions and their 2D contour plot, for N=6, are given in Figure 15 and Figure 16, respectively.

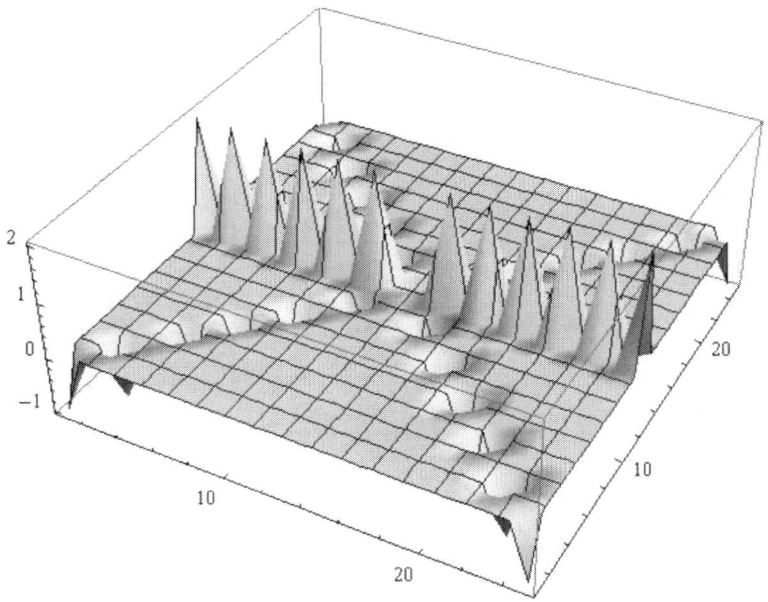

Figure 15. Coefficients of impulse response h2(N=6,r,k) of the second novel class of the 2D FIR filter functions.

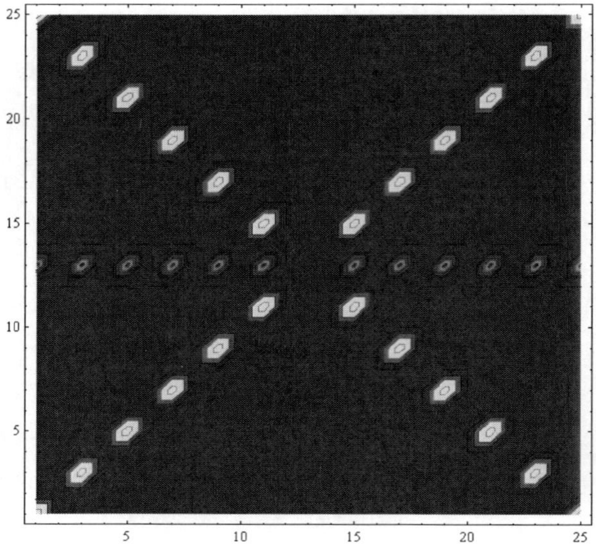

Figure 16. 2D Contour plot of coefficients of impulse response h2(N=6,r,k) of the second novel class of the 2D FIR filter functions.

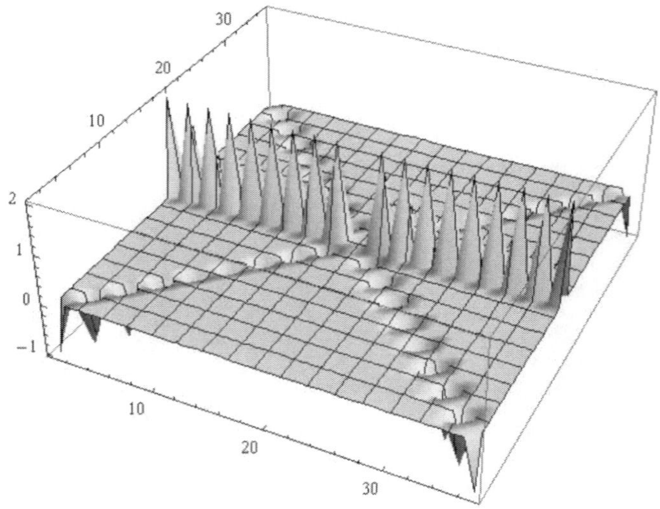

Figure 17. Coefficients of impulse response h2(N=9,r,k) of the second novel class of the 2D FIR filter functions.

The values of coefficients of impulse response h2(N=6,r,k) of the second novel class of the 2D FIR filter functions are given below:

h2(N=6,r,k)=
{{1,0,0,0,0,0,0,0,0,0,0,0,0,0,0,0,0,0,0,0,0,0,0,0,1},
{0,0,0,0,0,0,0,0,0,0,0,0,0,0,0,0,0,0,0,0,0,0,0,0,0},
{0,0,1,0,0,0,0,0,0,0,0,0,0,0,0,0,0,0,0,0,0,0,1,0,0},
{0,0,0,0,0,0,0,0,0,0,0,0,0,0,0,0,0,0,0,0,0,0,0,0,0},
{0,0,0,0,1,0,0,0,0,0,0,0,0,0,0,0,0,0,0,0,1,0,0,0,0},
{0,0,0,0,0,0,0,0,0,0,0,0,0,0,0,0,0,0,0,0,0,0,0,0,0},
{0,0,0,0,0,0,1,0,0,0,0,0,0,0,0,0,0,0,1,0,0,0,0,0,0},
{0,0,0,0,0,0,0,0,0,0,0,0,0,0,0,0,0,0,0,0,0,0,0,0,0},
{0,0,0,0,0,0,0,0,1,0,0,0,0,0,0,0,1,0,0,0,0,0,0,0,0},
{0,0,0,0,0,0,0,0,0,0,0,0,0,0,0,0,0,0,0,0,0,0,0,0,0},
{0,0,0,0,0,0,0,0,0,0,1,0,0,0,1,0,0,0,0,0,0,0,0,0,0},
{0,0,0,0,0,0,0,0,0,0,0,0,0,0,0,0,0,0,0,0,0,0,0,0,0},
{-2,0,-2,0,-2,0,-2,0,-2,0,-2,0,0,0,-2,0,-2,0,-2,0,-2,0,-2,0,-2},
{0,0,0,0,0,0,0,0,0,0,0,0,0,0,0,0,0,0,0,0,0,0,0,0,0},
{0,0,0,0,0,0,0,0,0,0,1,0,0,0,1,0,0,0,0,0,0,0,0,0,0},
{0,0,0,0,0,0,0,0,0,0,0,0,0,0,0,0,0,0,0,0,0,0,0,0,0},
{0,0,0,0,0,0,0,0,1,0,0,0,0,0,0,0,1,0,0,0,0,0,0,0,0},
{0,0,0,0,0,0,0,0,0,0,0,0,0,0,0,0,0,0,0,0,0,0,0,0,0},
{0,0,0,0,0,0,1,0,0,0,0,0,0,0,0,0,0,0,1,0,0,0,0,0,0},
{0,0,0,0,0,0,0,0,0,0,0,0,0,0,0,0,0,0,0,0,0,0,0,0,0},
{0,0,0,0,1,0,0,0,0,0,0,0,0,0,0,0,0,0,0,0,1,0,0,0,0},
{0,0,0,0,0,0,0,0,0,0,0,0,0,0,0,0,0,0,0,0,0,0,0,0,0},
{0,0,1,0,0,0,0,0,0,0,0,0,0,0,0,0,0,0,0,0,0,0,1,0,0},
{0,0,0,0,0,0,0,0,0,0,0,0,0,0,0,0,0,0,0,0,0,0,0,0,0},
{1,0,0,0,0,0,0,0,0,0,0,0,0,0,0,0,0,0,0,0,0,0,0,0,1}}. (40)

The frequency response characteristics in the z domain for N=9 is:

$$\begin{aligned}
H_2(N=9,z_1,z_2) &= 1+z_1^{-36}-2z_1^{-18}+z_2^{-36}+z_1^{-36}z_2^{-36}-2z_1^{-18}z_2^{-36}+z_1^{-34}z_2^{-34}- \\
&2z_1^{-18}z_2^{-34}+z_1^{-2}z_2^{-34}+z_1^{-32}z_2^{-32}-2z_1^{-18}z_2^{-32}+z_1^{-4}z_2^{-32}+z_1^{-30}z_2^{-30}-2z_1^{-18}z_2^{-30}+ \\
&z_1^{-6}z_2^{-30}+z_1^{-28}z_2^{-28}-2z_1^{-18}z_2^{-28}+z_1^{-8}z_2^{-28}+z_1^{-26}z_2^{-26}-2z_1^{-18}z_2^{-26}+z_1^{-10}z_2^{-26}+ \\
&z_1^{-24}z_2^{-24}-2z_1^{-18}z_2^{-24}+z_1^{-12}z_2^{-24}+z_1^{-22}z_2^{-22}-2z_1^{-18}z_2^{-22}+z_1^{-14}z_2^{-22}+z_1^{-20}z_2^{-20}- \\
&2z_1^{-18}z_2^{-20}+z_1^{-16}z_2^{-20}+z_1^{-20}z_2^{-16}-2z_1^{-18}z_2^{-16}+z_1^{-16}z_2^{-16}+z_1^{-22}z_2^{-14}-2z_1^{-18}z_2^{-14}+ \\
&z_1^{-14}z_2^{-14}+z_1^{-24}z_2^{-12}-2z_1^{-18}z_2^{-12}+z_1^{-12}z_2^{-12}+z_1^{-26}z_2^{-10}-2z_1^{-18}z_2^{-10}+z_1^{-10}z_2^{-10}+ \\
&z_1^{-28}z_2^{-8}-2z_1^{-18}z_2^{-8}+z_1^{-8}z_2^{-8}+z_1^{-30}z_2^{-6}-2z_1^{-18}z_2^{-6}+z_1^{-6}z_2^{-6}+z_1^{-32}z_2^{-4}- \\
&2z_1^{-18}z_2^{-4}+z_1^{-4}z_2^{-4}+z_1^{-34}z_2^{-2}-2z_1^{-18}z_2^{-2}+z_1^{-2}z_2^{-2}
\end{aligned} \tag{41}$$

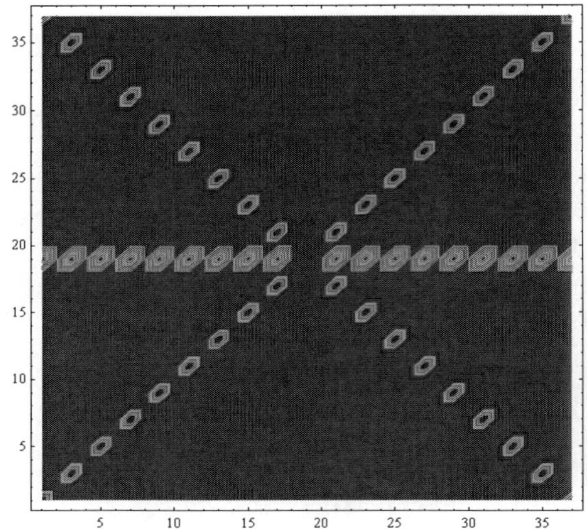

Figure 18. 2D contour plot of coefficients of impulse response h2(N=9,r,k) of the second novel class of the 2D FIR filter functions.

The coefficients of impulse response of the second novel class of the 2D FIR filter functions and their 2D contour plot, for N=9, are given in Figures 17 and18, respectively. The values of coefficients of impulse response h2(N=9,r,k) of the second novel class of the 2D FIR filter function are given below:

h2(N=9,r,k) =
{{1,0,0,0,0,0,0,0,0,0,0,0,0,0,0,0,0,0,0,0,0,0,0,0,0,0,0,0,0,0,0,0,0,0,0,0,1},
{0,0,0,0,0,0,0,0,0,0,0,0,0,0,0,0,0,0,0,0,0,0,0,0,0,0,0,0,0,0,0,0,0,0,0,0,0},
{0,0,1,0,0,0,0,0,0,0,0,0,0,0,0,0,0,0,0,0,0,0,0,0,0,0,0,0,0,0,0,0,0,0,1,0,0},
{0,0,0,0,0,0,0,0,0,0,0,0,0,0,0,0,0,0,0,0,0,0,0,0,0,0,0,0,0,0,0,0,0,0,0,0,0},
{0,0,0,0,1,0,0,0,0,0,0,0,0,0,0,0,0,0,0,0,0,0,0,0,0,0,0,0,0,0,0,0,1,0,0,0,0},
{0,0,0,0,0,0,0,0,0,0,0,0,0,0,0,0,0,0,0,0,0,0,0,0,0,0,0,0,0,0,0,0,0,0,0,0,0},
{0,0,0,0,0,0,1,0,0,0,0,0,0,0,0,0,0,0,0,0,0,0,0,0,0,0,0,0,0,0,1,0,0,0,0,0,0},
{0,0,0,0,0,0,0,0,0,0,0,0,0,0,0,0,0,0,0,0,0,0,0,0,0,0,0,0,0,0,0,0,0,0,0,0,0},
{0,0,0,0,0,0,0,0,1,0,0,0,0,0,0,0,0,0,0,0,0,0,0,0,0,0,0,0,1,0,0,0,0,0,0,0,0},
{0,0,0,0,0,0,0,0,0,0,0,0,0,0,0,0,0,0,0,0,0,0,0,0,0,0,0,0,0,0,0,0,0,0,0,0,0},
{0,0,0,0,0,0,0,0,0,0,1,0,0,0,0,0,0,0,0,0,0,0,0,0,0,0,1,0,0,0,0,0,0,0,0,0,0},
{0,0,0,0,0,0,0,0,0,0,0,0,0,0,0,0,0,0,0,0,0,0,0,0,0,0,0,0,0,0,0,0,0,0,0,0,0},
{0,0,0,0,0,0,0,0,0,0,0,0,1,0,0,0,0,0,0,0,0,0,0,0,1,0,0,0,0,0,0,0,0,0,0,0,0},
{0,0,0,0,0,0,0,0,0,0,0,0,0,0,0,0,0,0,0,0,0,0,0,0,0,0,0,0,0,0,0,0,0,0,0,0,0},
{0,0,0,0,0,0,0,0,0,0,0,0,0,0,1,0,0,0,0,0,0,0,1,0,0,0,0,0,0,0,0,0,0,0,0,0,0},

{0,0,0,0,0,0,0,0,0,0,0,0,0,0,0,0,0,0,0,0,0,0,0,0,0,0,0,0,0,0,0,0,0,0,0,0,0},
{0,0,0,0,0,0,0,0,0,0,0,0,0,0,0,0,1,0,0,0,1,0,0,0,0,0,0,0,0,0,0,0,0,0,0,0,0},
{0,0,0,0,0,0,0,0,0,0,0,0,0,0,0,0,0,0,0,0,0,0,0,0,0,0,0,0,0,0,0,0,0,0,0,0,0},
{-2,0,-2,0,-2,0,-2,0,-2,0,-2,0,-2,0,-2,0,-2,0,0,0,-2,0,-2,0,-2,0,-2,0,-2,0,-2,0,-
2,0,-2,0,-2},
{0,0,0,0,0,0,0,0,0,0,0,0,0,0,0,0,0,0,0,0,0,0,0,0,0,0,0,0,0,0,0,0,0,0,0,0,0},
{0,0,0,0,0,0,0,0,0,0,0,0,0,0,0,0,1,0,0,0,1,0,0,0,0,0,0,0,0,0,0,0,0,0,0,0,0},
{0,0,0,0,0,0,0,0,0,0,0,0,0,0,0,0,0,0,0,0,0,0,0,0,0,0,0,0,0,0,0,0,0,0,0,0,0},
{0,0,0,0,0,0,0,0,0,0,0,0,0,0,1,0,0,0,0,0,0,0,1,0,0,0,0,0,0,0,0,0,0,0,0,0,0},
{0,0,0,0,0,0,0,0,0,0,0,0,0,0,0,0,0,0,0,0,0,0,0,0,0,0,0,0,0,0,0,0,0,0,0,0,0},
{0,0,0,0,0,0,0,0,0,0,0,0,1,0,0,0,0,0,0,0,0,0,0,0,1,0,0,0,0,0,0,0,0,0,0,0,0},
{0,0,0,0,0,0,0,0,0,0,0,0,0,0,0,0,0,0,0,0,0,0,0,0,0,0,0,0,0,0,0,0,0,0,0,0,0},
{0,0,0,0,0,0,0,0,0,0,1,0,0,0,0,0,0,0,0,0,0,0,0,0,0,0,1,0,0,0,0,0,0,0,0,0,0},
{0,0,0,0,0,0,0,0,0,0,0,0,0,0,0,0,0,0,0,0,0,0,0,0,0,0,0,0,0,0,0,0,0,0,0,0,0},
{0,0,0,0,0,0,0,0,1,0,0,0,0,0,0,0,0,0,0,0,0,0,0,0,0,0,0,0,1,0,0,0,0,0,0,0,0},
{0,0,0,0,0,0,0,0,0,0,0,0,0,0,0,0,0,0,0,0,0,0,0,0,0,0,0,0,0,0,0,0,0,0,0,0,0},
{0,0,0,0,0,0,1,0,0,0,0,0,0,0,0,0,0,0,0,0,0,0,0,0,0,0,0,0,0,0,1,0,0,0,0,0,0},
{0,0,0,0,0,0,0,0,0,0,0,0,0,0,0,0,0,0,0,0,0,0,0,0,0,0,0,0,0,0,0,0,0,0,0,0,0},
{0,0,0,0,1,0,0,0,0,0,0,0,0,0,0,0,0,0,0,0,0,0,0,0,0,0,0,0,0,0,0,0,1,0,0,0,0},
{0,0,0,0,0,0,0,0,0,0,0,0,0,0,0,0,0,0,0,0,0,0,0,0,0,0,0,0,0,0,0,0,0,0,0,0,0},
{0,0,1,0,0,0,0,0,0,0,0,0,0,0,0,0,0,0,0,0,0,0,0,0,0,0,0,0,0,0,0,0,0,0,1,0,0},
{0,0,0,0,0,0,0,0,0,0,0,0,0,0,0,0,0,0,0,0,0,0,0,0,0,0,0,0,0,0,0,0,0,0,0,0,0},
{1,0,0,0,0,0,0,0,0,0,0,0,0,0,0,0,0,0,0,0,0,0,0,0,0,0,0,0,0,0,0,0,0,0,0,0,1}}

(42)

For the linear phase 2D symmetric FIR digital filters generated by the proposed technique, the following symmetries are valid:

$$H_2\left(N, z_1, -z_2\right) = H_2\left(N, z_1, z_2\right) \tag{43}$$

$$H_2\left(N, -z_1, z_2\right) = H_2\left(N, z_1, z_2\right) \tag{44}$$

The linear phase function of the 2D symmetric FIR digital filter defined by Eq. (33) has the following form for $z_1 \to e^{i\omega_1}$ and $z_2 \to e^{j\omega_2}$:

$$\varphi_2(N, \omega_1, \omega_2) = -i2N\omega_1 - j2N\omega_2. \tag{45}$$

Frequency response characteristics of the second novel class of the 2D FIR filter functions, $H_2(M,K,e^{i\omega_1},e^{j\omega_2})$, obtained after inserting Eq.(32) in Eq.(26), has the following form:

$$H_2(M,K,e^{i\omega_1},e^{j\omega_2}) = K_1\, e^{-i6MK\omega_1} e^{-j6MK\omega_2}$$

$$[\sum_{k=1}^{M-1}(\cos(2k\omega_1)-1)\cos(2k\omega_2)\sum_{r=1}^{M}(\cos(2r\omega_1)-1)\cos(2r\omega_2)$$

$$\sum_{m=1}^{M+1}(\cos(2m\omega_1)-1)\cos(2m\omega_2)]^K \tag{46}$$

Amplitude response characteristics, magnitude response characteristics and linear phase response characteristics of the second novel class of the 2D FIR filter functions, obtained from Eq.(46), are:

$$A_2(M,K,\omega_1,\omega_2) = [\sum_{k=1}^{M-1}(\cos(2k\omega_1)-1)\cos(2k\omega_2)$$

$$\sum_{r=1}^{M}(\cos(2r\omega_1)-1)\cos(2r\omega_2)\sum_{m=1}^{M+1}(\cos(2m\omega_1)-1)\cos(2m\omega_2)]^K \tag{47}$$

$$\left|H_2\left(M,K,e^{j\omega_1},e^{j\omega_2}\right)\right| = \left|A_2(M,K,\omega_1,\omega_2)\right| \tag{48}$$

$$\varphi_2(M,K,\omega_1,\omega_2) = -i6MK\omega_1 - j6MK\omega_2 \tag{49}$$

respectively.

5. The Third Novel Class of Non-Recursive 2D FIR Filter Functions

We propose the third novel class of non-causal linear phase and non-recursive 2D symmetric FIR filter functions with dimension $N*N$ in the following explicit form:

$$H_3\left(N, z_1, z_2\right) = K_1 \sum_{r=1}^{N} (z_1^{+2r} + 2 + z_1^{-2r}) \times (z_2^{+2r} + z_2^{-2r}) \tag{50}$$

The third novel class of causal 2D FIR filter function can be generated by simple multiplication of Eq.(50) with factor $z_1^{-2N} * z_2^{-2N}$ as:

$$H_3(N, z_1, z_2) = K_1 \sum_{r=1}^{N} (z_1^{-2N+2r} + 2z_1^{-2N} + z_1^{-2N-2r}) \times (z_2^{-2N+2r} + z_2^{-2N-2r}) \tag{51}$$

Frequency response characteristics, amplitude response characteristics, magnitude response characteristics and linear phase response characteristics of the third novel class of 2D symmetric FIR filter functions are given by the following equations:

$$H_3(N, e^{j\omega_1}, e^{j\omega_2}) = K_1\, e^{i2N\omega_1} e^{j2N\omega_2} \sum_{r=1}^{N} (\cos(2r\omega_1) + 1)\cos(2r\omega_2) \tag{52}$$

$$A_3\left(N, \omega_1, \omega_2\right) = \sum_{r=1}^{N} (\cos(2r\omega_1) + 1)\ \cos(2r\omega_2) \tag{53}$$

$$\left|H_3\left(N, e^{i\omega_1}, e^{j\omega_2}\right)\right| = \left|A_3(N, \omega_1, \omega_2)\right| \tag{54}$$

$$\varphi_3(N, \omega_1, \omega_2) = -\, i2N\omega_1 - j2N\omega_2 \tag{55}$$

respectively.

The frequency response characteristics in the z domain for N=5 is:

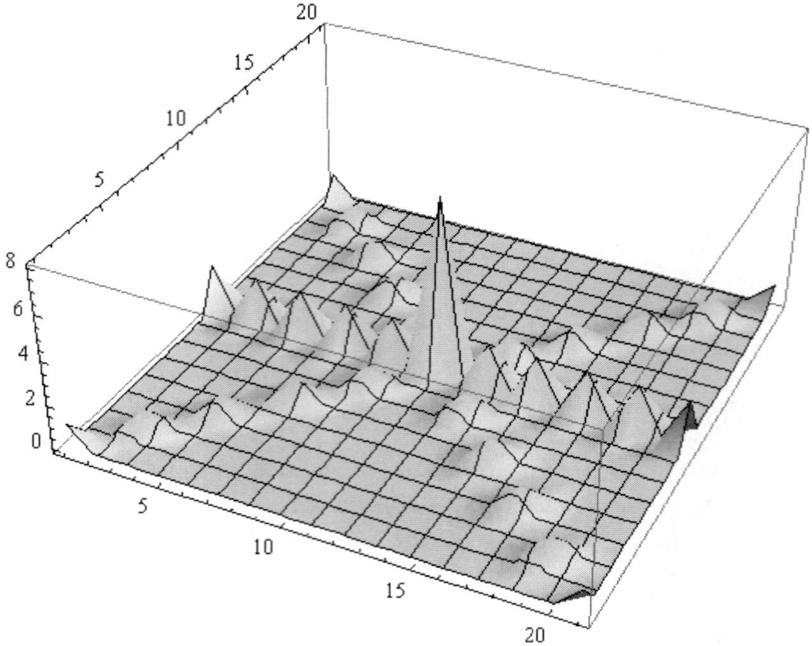

Figure 19. Coefficients of impulse response h3(N=5,r,k) of the third novel class of the 2D FIR filter functions.

$$\begin{aligned}
H_3(N=5,z_1,z_2) &= 1+z_1^{-20}+2z_1^{-10}+z_2^{-20}+z_1^{-20}z_2^{-20}+2z_1^{-10}z_2^{-20}+z_1^{-18}z_2^{-18}+\\
&2z_1^{-10}z_2^{-18}+z_1^{-2}z_2^{-18}+z_1^{-16}z_2^{-16}+2z_1^{-10}z_2^{-16}+z_1^{-4}z_2^{-16}+z_1^{-14}z_2^{-14}+2z_1^{-10}z_2^{-14}+\\
&z_1^{-6}z_2^{-14}+z_1^{-12}z_2^{-12}+2z_1^{-10}z_2^{-12}+z_1^{-8}z_2^{-12}+8z_1^{-10}z_2^{-10}+z_1^{-12}z_2^{-8}+2z_1^{-10}z_2^{-8}+\\
&z_1^{-8}z_2^{-8}+z_1^{-14}z_2^{-6}+2z_1^{-10}z_2^{-6}+z_1^{-6}z_2^{-6}+z_1^{-16}z_2^{-4}+2z_1^{-10}z_2^{-4}+z_1^{-4}z_2^{-4}+\\
&z_1^{-18}z_2^{-2}+2z_1^{-10}z_2^{-2}+z_1^{-2}z_2^{-2}
\end{aligned} \tag{56}$$

The coefficients of impulse response of the third novel class of the 2D FIR filter functions and their 2D contour plot, for N=5, are given in Figure 19 and Figure 20, respectively.

The values of coefficients of impulse response h3(N=5,r,k) of the third novel class of the 2D FIR filter functions are given below:

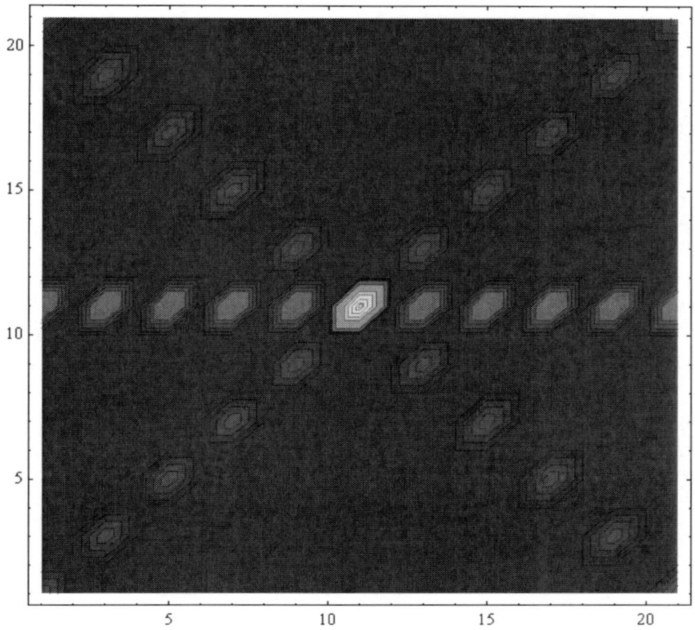

Figure 20. 2D contour plot of coefficients of impulse response h3(N=5,r,k) of the third novel class of the 2D FIR filter functions.

h3(N=5,r,k)=
{{1,0,0,0,0,0,0,0,0,0,0,0,0,0,0,0,0,0,0,0,1},
{0,0,0,0,0,0,0,0,0,0,0,0,0,0,0,0,0,0,0,0,0},
{0,0,1,0,0,0,0,0,0,0,0,0,0,0,0,0,0,0,1,0,0},
{0,0,0,0,0,0,0,0,0,0,0,0,0,0,0,0,0,0,0,0,0},
{0,0,0,0,1,0,0,0,0,0,0,0,0,0,0,0,1,0,0,0,0},
{0,0,0,0,0,0,0,0,0,0,0,0,0,0,0,0,0,0,0,0,0},
{0,0,0,0,0,0,1,0,0,0,0,0,0,0,1,0,0,0,0,0,0},
{0,0,0,0,0,0,0,0,0,0,0,0,0,0,0,0,0,0,0,0,0},
{0,0,0,0,0,0,0,0,1,0,0,0,1,0,0,0,0,0,0,0,0},
{0,0,0,0,0,0,0,0,0,0,0,0,0,0,0,0,0,0,0,0,0},
{2,0,2,0,2,0,2,0,2,0,8,0,2,0,2,0,2,0,2,0,2},
{0,0,0,0,0,0,0,0,0,0,0,0,0,0,0,0,0,0,0,0,0},
{0,0,0,0,0,0,0,0,1,0,0,0,1,0,0,0,0,0,0,0,0},
{0,0,0,0,0,0,0,0,0,0,0,0,0,0,0,0,0,0,0,0,0},
{0,0,0,0,0,0,1,0,0,0,0,0,0,0,1,0,0,0,0,0,0},
{0,0,0,0,0,0,0,0,0,0,0,0,0,0,0,0,0,0,0,0,0},
{0,0,0,0,1,0,0,0,0,0,0,0,0,0,0,0,1,0,0,0,0},

$$\{0,0,0,0,0,0,0,0,0,0,0,0,0,0,0,0,0,0,0,0,0,0,0,0,0\},$$
$$\{0,0,1,0,0,0,0,0,0,0,0,0,0,0,0,0,0,0,0,0,0,0,1,0,0\},$$
$$\{0,0,0,0,0,0,0,0,0,0,0,0,0,0,0,0,0,0,0,0,0,0,0,0,0\},$$
$$\{1,0,0,0,0,0,0,0,0,0,0,0,0,0,0,0,0,0,0,0,0,0,0,0,1\}\}. \quad (57)$$

The frequency response characteristics in the z domain for N=6 is:

$$H_3(N=6, z_1, z_2) = 1 + z_1^{-24} + 2z_1^{-12} + z_2^{-24} + z_1^{-24}z_2^{-24} + 2z_1^{-12}z_2^{-24} + z_1^{-22}z_2^{-22} + 2z_1^{-12}z_2^{-22} + z_1^{-2}z_2^{-22} + z_1^{-20}z_2^{-20} + 2z_1^{-12}z_2^{-20} + z_1^{-4}z_2^{-20} + z_1^{-18}z_2^{-18} + 2z_1^{-12}z_2^{-18} + z_1^{-6}z_2^{-18} + z_1^{-16}z_2^{-16} + 2z_1^{-12}z_2^{-16} + z_1^{-8}z_2^{-16} + z_1^{-14}z_2^{-14} + 2z_1^{-12}z_2^{-14} + z_1^{-10}z_2^{-14} + 8z_1^{-12}z_2^{-12} + z_1^{-14}z_2^{-10} + 2z_1^{-12}z_2^{-10} + z_1^{-10}z_2^{-10} + z_1^{-16}z_2^{-8} + 2z_1^{-12}z_2^{-8} + z_1^{-8}z_2^{-8} + z_1^{-18}z_2^{-6} + 2z_1^{-12}z_2^{-6} + z_1^{-6}z_2^{-6} + z_1^{-20}z_2^{-4} + 2z_1^{-12}z_2^{-4} + z_1^{-4}z_2^{-4} + z_1^{-22}z_2^{-2} + 2z_1^{-12}z_2^{-2} + z_1^{-2}z_2^{-2} \quad (58)$$

The coefficients of impulse response of the third novel class of the 2D FIR filter functions and their 2D contour plot, for N=6, are given in Figure 21 and Figure 22, respectively.

The values of coefficients of impulse response h3(N=6,r,k) of the third novel class of the 2D FIR filter functions are given below:

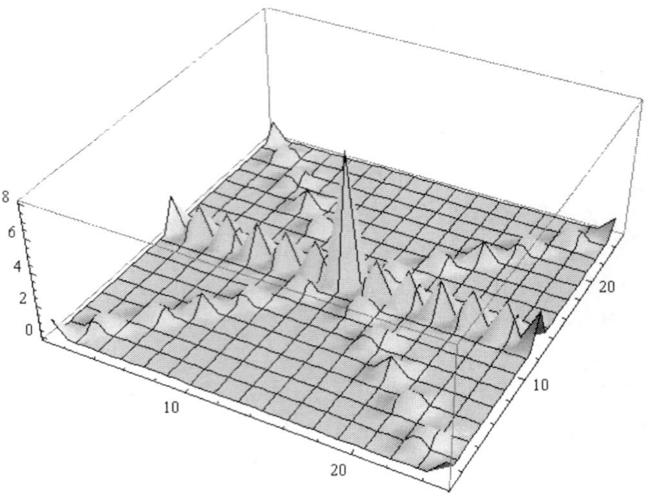

Figure 21. Coefficients of impulse response h3(N=6,r,k) of the third novel class of the 2D FIR filter functions.

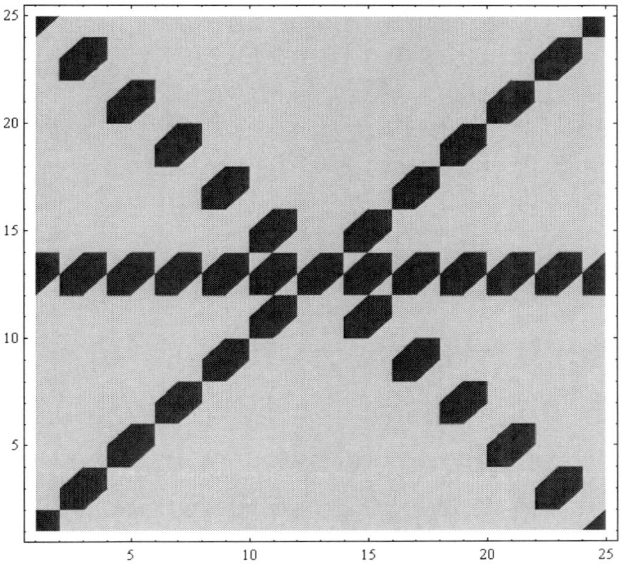

Figure 22. 2D contour plot of coefficients of impulse response h3(N=6,r,k) of the third novel class of the 2D FIR filter functions.

h3(N=6,r,k) =
{{1,0,0,0,0,0,0,0,0,0,0,0,0,0,0,0,0,0,0,0,0,0,0,0,1},
{0,0,0,0,0,0,0,0,0,0,0,0,0,0,0,0,0,0,0,0,0,0,0,0,0},
{0,0,1,0,0,0,0,0,0,0,0,0,0,0,0,0,0,0,0,0,0,0,1,0,0},
{0,0,0,0,0,0,0,0,0,0,0,0,0,0,0,0,0,0,0,0,0,0,0,0,0},
{0,0,0,0,1,0,0,0,0,0,0,0,0,0,0,0,0,0,0,0,1,0,0,0,0},
{0,0,0,0,0,0,0,0,0,0,0,0,0,0,0,0,0,0,0,0,0,0,0,0,0},
{0,0,0,0,0,0,1,0,0,0,0,0,0,0,0,0,0,0,1,0,0,0,0,0,0},
{0,0,0,0,0,0,0,0,0,0,0,0,0,0,0,0,0,0,0,0,0,0,0,0,0},
{0,0,0,0,0,0,0,0,1,0,0,0,0,0,0,0,1,0,0,0,0,0,0,0,0},
{0,0,0,0,0,0,0,0,0,0,0,0,0,0,0,0,0,0,0,0,0,0,0,0,0},
{0,0,0,0,0,0,0,0,0,0,1,0,0,0,1,0,0,0,0,0,0,0,0,0,0},
{0,0,0,0,0,0,0,0,0,0,0,0,0,0,0,0,0,0,0,0,0,0,0,0,0},
{2,0,2,0,2,0,2,0,2,0,2,0,8,0,2,0,2,0,2,0,2,0,2,0,2},
{0,0,0,0,0,0,0,0,0,0,0,0,0,0,0,0,0,0,0,0,0,0,0,0,0},
{0,0,0,0,0,0,0,0,0,0,1,0,0,0,1,0,0,0,0,0,0,0,0,0,0},
{0,0,0,0,0,0,0,0,0,0,0,0,0,0,0,0,0,0,0,0,0,0,0,0,0},
{0,0,0,0,0,0,0,0,1,0,0,0,0,0,0,0,1,0,0,0,0,0,0,0,0},
{0,0,0,0,0,0,0,0,0,0,0,0,0,0,0,0,0,0,0,0,0,0,0,0,0},
{0,0,0,0,0,0,1,0,0,0,0,0,0,0,0,0,0,0,1,0,0,0,0,0,0},

{0,0,0,0,0,0,0,0,0,0,0,0,0,0,0,0,0,0,0,0,0,0,0,0,0,0,0},
{0,0,0,0,1,0,0,0,0,0,0,0,0,0,0,0,0,0,0,0,0,0,1,0,0,0,0},
{0,0,0,0,0,0,0,0,0,0,0,0,0,0,0,0,0,0,0,0,0,0,0,0,0,0,0},
{0,0,1,0,0,0,0,0,0,0,0,0,0,0,0,0,0,0,0,0,0,0,0,0,1,0,0},
{0,0,0,0,0,0,0,0,0,0,0,0,0,0,0,0,0,0,0,0,0,0,0,0,0,0,0},
{1,0,0,0,0,0,0,0,0,0,0,0,0,0,0,0,0,0,0,0,0,0,0,0,0,0,1}}. (59)

The frequency response characteristics in the z domain for N=9 is:

$$\begin{aligned}
&H_3(N=9, z_1, z_2) = 1 + z_1^{-36} + 2z_1^{-18} + z_2^{-36} + z_1^{-36} z_2^{-36} + 2z_1^{-18} z_2^{-36} + z_1^{-34} z_2^{-34} + \\
&2z_1^{-18} z_2^{-34} + z_1^{-2} z_2^{-34} + z_1^{-32} z_2^{-32} + 2z_1^{-18} z_2^{-32} + z_1^{-4} z_2^{-32} + z_1^{-30} z_2^{-30} + 2z_1^{-18} z_2^{-30} + \\
&z_1^{-6} z_2^{-30} + z_1^{-28} z_2^{-28} + 2z_1^{-18} z_2^{-28} + z_1^{-8} z_2^{-28} + z_1^{-26} z_2^{-26} + 2z_1^{-18} z_2^{-26} + z_1^{-10} z_2^{-26} + \\
&z_1^{-24} z_2^{-24} + 2z_1^{-18} z_2^{-24} + z_1^{-12} z_2^{-24} + z_1^{-22} z_2^{-22} + 2z_1^{-18} z_2^{-22} + z_1^{-14} z_2^{-22} + z_1^{-20} z_2^{-20} + \\
&2z_1^{-18} z_2^{-20} + z_1^{-16} z_2^{-20} + 8z_1^{-18} z_2^{-18} + z_1^{-20} z_2^{-16} + 2z_1^{-18} z_2^{-16} + z_1^{-16} z_2^{-16} + z_1^{-22} z_2^{-14} + \\
&2z_1^{-18} z_2^{-14} + z_1^{-14} z_2^{-14} + z_1^{-24} z_2^{-12} + 2z_1^{-18} z_2^{-12} + z_1^{-12} z_2^{-12} + z_1^{-26} z_2^{-10} + 2z_1^{-18} z_2^{-10} + \\
&z_1^{-10} z_2^{-10} + z_1^{-28} z_2^{-8} + 2z_1^{-18} z_2^{-8} + z_1^{-8} z_2^{-8} + z_1^{-30} z_2^{-6} + 2z_1^{-18} z_2^{-6} + z_1^{-6} z_2^{-6} + \\
&z_1^{-32} z_2^{-4} + 2z_1^{-18} z_2^{-4} + z_1^{-4} z_2^{-4} + z_1^{-34} z_2^{-2} + 2z_1^{-18} z_2^{-2} + z_1^{-2} z_2^{-2}
\end{aligned} \tag{60}$$

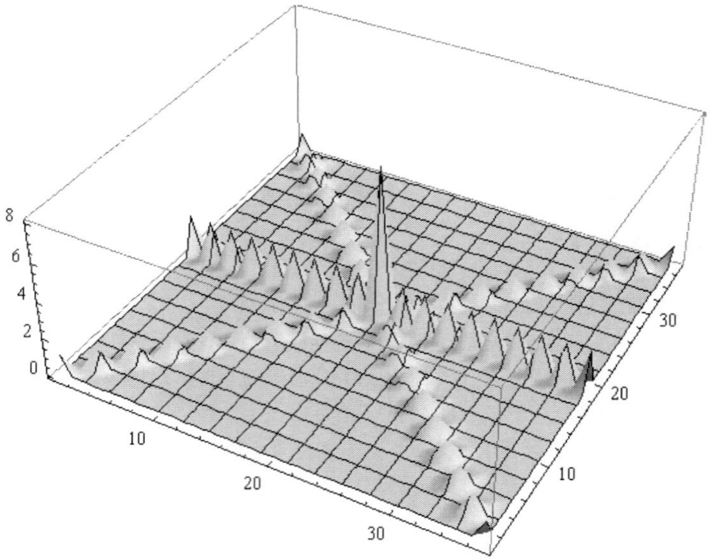

Figure 23. Coefficients of impulse response h3(N=9,r,k) of the third novel class of the 2D FIR filter functions.

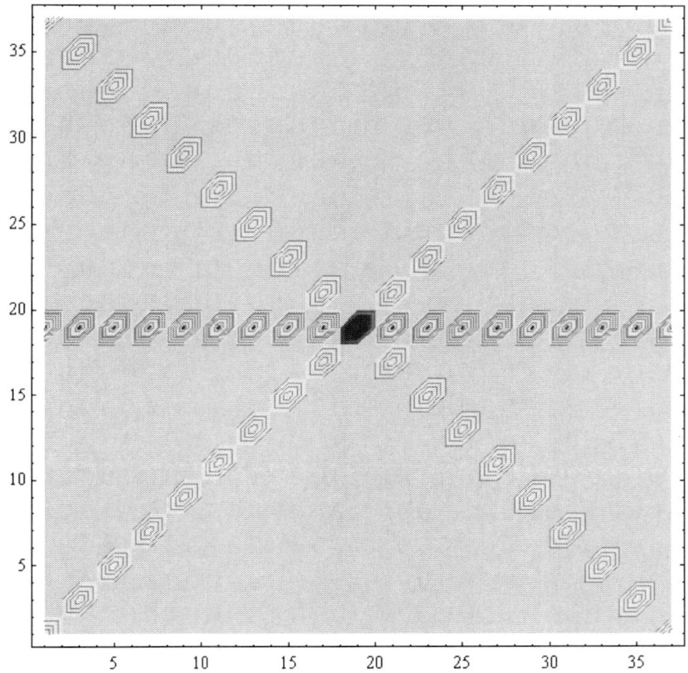

Figure 24. Contour plot of coefficients of impulse response h3(N=9,r,k) of the new third novel class of the 2D FIR filter functions.

The coefficients of impulse response of the third novel class of the 2D FIR filter function and their 2D contour plot, for N=9, are given in Figure 23 and Figure 24, respectively.

The values of coefficients of impulse response h3(N=9,r,k) of the new third class of the 2D FIR filter function are given below:

h3(N=9,r,k)=
{{1,0,0,0,0,0,0,0,0,0,0,0,0,0,0,0,0,0,0,0,0,0,0,0,0,0,0,0,0,0,0,0,0,0,0,0,1},
{0,0,0,0,0,0,0,0,0,0,0,0,0,0,0,0,0,0,0,0,0,0,0,0,0,0,0,0,0,0,0,0,0,0,0,0,0},
{0,0,1,0,0,0,0,0,0,0,0,0,0,0,0,0,0,0,0,0,0,0,0,0,0,0,0,0,0,0,0,0,0,0,1,0,0},
{0,0,0,0,0,0,0,0,0,0,0,0,0,0,0,0,0,0,0,0,0,0,0,0,0,0,0,0,0,0,0,0,0,0,0,0,0},
{0,0,0,0,1,0,0,0,0,0,0,0,0,0,0,0,0,0,0,0,0,0,0,0,0,0,0,0,0,0,0,0,1,0,0,0,0},
{0,0,0,0,0,0,0,0,0,0,0,0,0,0,0,0,0,0,0,0,0,0,0,0,0,0,0,0,0,0,0,0,0,0,0,0,0},
{0,0,0,0,0,0,1,0,0,0,0,0,0,0,0,0,0,0,0,0,0,0,0,0,0,0,0,0,0,0,1,0,0,0,0,0,0},
{0,0,0,0,0,0,0,0,0,0,0,0,0,0,0,0,0,0,0,0,0,0,0,0,0,0,0,0,0,0,0,0,0,0,0,0,0},
{0,0,0,0,0,0,0,0,1,0,0,0,0,0,0,0,0,0,0,0,0,0,0,0,0,0,0,0,1,0,0,0,0,0,0,0,0},

{0,0,0,0,0,0,0,0,0,0,0,0,0,0,0,0,0,0,0,0,0,0,0,0,0,0,0,0,0,0,0,0,0,0,0,0,0},
{0,0,0,0,0,0,0,0,0,0,1,0,0,0,0,0,0,0,0,0,0,0,0,0,0,0,1,0,0,0,0,0,0,0,0,0,0},
{0,0,0,0,0,0,0,0,0,0,0,0,0,0,0,0,0,0,0,0,0,0,0,0,0,0,0,0,0,0,0,0,0,0,0,0,0},
{0,0,0,0,0,0,0,0,0,0,0,0,1,0,0,0,0,0,0,0,0,0,0,0,1,0,0,0,0,0,0,0,0,0,0,0,0},
{0,0,0,0,0,0,0,0,0,0,0,0,0,0,0,0,0,0,0,0,0,0,0,0,0,0,0,0,0,0,0,0,0,0,0,0,0},
{0,0,0,0,0,0,0,0,0,0,0,0,0,0,1,0,0,0,0,0,0,0,1,0,0,0,0,0,0,0,0,0,0,0,0,0,0},
{0,0,0,0,0,0,0,0,0,0,0,0,0,0,0,0,0,0,0,0,0,0,0,0,0,0,0,0,0,0,0,0,0,0,0,0,0},
{0,0,0,0,0,0,0,0,0,0,0,0,0,0,0,0,1,0,0,0,1,0,0,0,0,0,0,0,0,0,0,0,0,0,0,0,0},
{0,0,0,0,0,0,0,0,0,0,0,0,0,0,0,0,0,0,0,0,0,0,0,0,0,0,0,0,0,0,0,0,0,0,0,0,0},
{2,0,2,0,2,0,2,0,2,0,2,0,2,0,2,0,2,0,8,0,2,0,2,0,2,0,2,0,2,0,2,0,2,0,2,0,2},
{0,0,0,0,0,0,0,0,0,0,0,0,0,0,0,0,0,0,0,0,0,0,0,0,0,0,0,0,0,0,0,0,0,0,0,0,0},
{0,0,0,0,0,0,0,0,0,0,0,0,0,0,0,0,1,0,0,0,1,0,0,0,0,0,0,0,0,0,0,0,0,0,0,0,0},
{0,0,0,0,0,0,0,0,0,0,0,0,0,0,0,0,0,0,0,0,0,0,0,0,0,0,0,0,0,0,0,0,0,0,0,0,0},
{0,0,0,0,0,0,0,0,0,0,0,0,0,0,1,0,0,0,0,0,0,0,1,0,0,0,0,0,0,0,0,0,0,0,0,0,0},
{0,0,0,0,0,0,0,0,0,0,0,0,0,0,0,0,0,0,0,0,0,0,0,0,0,0,0,0,0,0,0,0,0,0,0,0,0},
{0,0,0,0,0,0,0,0,0,0,0,0,1,0,0,0,0,0,0,0,0,0,0,0,1,0,0,0,0,0,0,0,0,0,0,0,0},
{0,0,0,0,0,0,0,0,0,0,0,0,0,0,0,0,0,0,0,0,0,0,0,0,0,0,0,0,0,0,0,0,0,0,0,0,0},
{0,0,0,0,0,0,0,0,0,0,1,0,0,0,0,0,0,0,0,0,0,0,0,0,0,0,1,0,0,0,0,0,0,0,0,0,0},
{0,0,0,0,0,0,0,0,0,0,0,0,0,0,0,0,0,0,0,0,0,0,0,0,0,0,0,0,0,0,0,0,0,0,0,0,0},
{0,0,0,0,0,0,0,0,1,0,0,0,0,0,0,0,0,0,0,0,0,0,0,0,0,0,0,0,1,0,0,0,0,0,0,0,0},
{0,0,0,0,0,0,0,0,0,0,0,0,0,0,0,0,0,0,0,0,0,0,0,0,0,0,0,0,0,0,0,0,0,0,0,0,0},
{0,0,0,0,0,0,1,0,0,0,0,0,0,0,0,0,0,0,0,0,0,0,0,0,0,0,0,0,0,0,1,0,0,0,0,0,0},
{0,0,0,0,0,0,0,0,0,0,0,0,0,0,0,0,0,0,0,0,0,0,0,0,0,0,0,0,0,0,0,0,0,0,0,0,0},
{0,0,0,0,1,0,0,0,0,0,0,0,0,0,0,0,0,0,0,0,0,0,0,0,0,0,0,0,0,0,0,0,1,0,0,0,0},
{0,0,0,0,0,0,0,0,0,0,0,0,0,0,0,0,0,0,0,0,0,0,0,0,0,0,0,0,0,0,0,0,0,0,0,0,0},
{0,0,1,0,0,0,0,0,0,0,0,0,0,0,0,0,0,0,0,0,0,0,0,0,0,0,0,0,0,0,0,0,0,0,1,0,0},
{0,0,0,0,0,0,0,0,0,0,0,0,0,0,0,0,0,0,0,0,0,0,0,0,0,0,0,0,0,0,0,0,0,0,0,0,0},
{1,0,0,0,0,0,0,0,0,0,0,0,0,0,0,0,0,0,0,0,0,0,0,0,0,0,0,0,0,0,0,0,0,0,0,0,1}} .

(61)

For the linear phase 2D symmetric FIR digital filters generated by the proposed approximation technique, the following symmetries are valid:

$$H_3\left(N, z_1, -z_2\right) = H_3\left(N, z_1, z_2\right) \tag{62}$$

$$H_3(N,-z_1, z_2) = H_3(N, z_1, z_2) \tag{63}$$

Frequency response characteristics of the third novel class of the 2D FIR filter functions, $H_3(M,K,e^{i\omega_1},e^{j\omega_2})$, obtained after inserting Eq.(51) in Eq.(26), has the following form:

$$H_3(M,K,e^{i\omega_1},e^{j\omega_2}) = K_1\, e^{-i6MK\omega_1} e^{-j6MK\omega_2}$$
$$[(\sum_{k=1}^{M-1}(\cos(2k\omega_1)+1)\cos(2k\omega_2)\sum_{r=1}^{M}(\cos(2r\omega_1)+1)\cos(2r\omega_2)$$
$$\sum_{m=1}^{M+1}(\cos(2m\omega_1)+1)\cos(2m\omega_2)]^{K} \tag{64}$$

Amplitude response characteristics, magnitude response characteristics and linear phase response characteristics of the third novel class of the 2D FIR filter functions, obtained from Eq.(64), are:

$$A_3(M,K,\omega_1,\omega_2) = [\sum_{k=1}^{M-1}(\cos(2k\omega_1)+1)\cos(2k\omega_2)$$
$$\sum_{r=1}^{M}(\cos(2r\omega_1)+1)\cos(2r\omega_2)\sum_{m=1}^{M+1}(\cos(2m\omega_1)+1)\cos(2m\omega_2)]^{K} \tag{65}$$

$$\left|H_3(M,K,e^{j\omega_1},e^{j\omega_2})\right| = \left|A_3(M,K,\omega_1,\omega_2)\right| \tag{66}$$

$$\varphi_3(M,K,\omega_1,\omega_2) = -i6MK\omega_1 - j6MK\omega_2 \tag{67}$$

respectively.

6. Examples of the First Novel Class of Linear Phase and Non-Recursive 2D FIR Filter Functions

Examples of a proposed first class of 2D FIR filter functions obtained for particular values of free real integer parameters K and M from Eq. (10), applying Eq. (26) are given next in the chapter. The limitations of this proposed class 2D FIR filter functions are the same as for CIC filters because the values of free real integer parameters are discrete. Values of free real integer parameters are of arbitrary parity.

Independently of parameters parity, the proposed function has only filter property. The type of filter function for different parity of free real integer parameters is given in Table 1.

Proposed first novel class of 2D FIR filter function has limitations as it is only of symmetrical type (NxN) and the same limitations exist as for CIC filters for the filter order defined by the choice of free real integer parameters K and M [17-19, 22]. This means that the possibility to form a shape of linear-phase characteristics is limited.

Set of some possible explicit forms of linear-phase response characteristics for all three classes of 2D FIR filter functions, obtained for particular values of parameters M and K, are given in Table 2.

Table 1. Type of the filter for different parity of *K* and *M*

K	M	Type
Odd	Odd	Filter
Even	Odd	Filter
Odd	Even	Filter
Even	Even	Filter

Authors of this chapter would like to point out that the linear-phase response characteristics of proposed classes of 2D FIR filter functions can be easily determined as for 1D and 2D CIC filter functions and that the same limitations for the value of constant group delay exist as in the case of CIC filters.

Example 1:

We consider the example of the proposed first novel class of 2D FIR filter for odd M and K, M=7 and K=3. Phase response characteristics from Eq.(30) has the form:

Table 2. Some possible explicit forms of linear phase characteristics for particular values of free real integer parameters *K* and *M*

K	M	$\varphi_1(M,K,\omega_1,\omega_2) = -i6MK\omega_1 - j6MK\omega_2$	Type
3	4	$\varphi_1(4,3,\omega_1,\omega_2) = -i72\omega_1 - j72\omega_2$	Filter
3	5	$\varphi_1(5,3,\omega_1,\omega_2) = -i90\omega_1 - j90\omega_2$	Filter
3	6	$\varphi_1(6,3,\omega_1,\omega_2) = -i108\omega_1 - j108\omega_2$	Filter
3	7	$\varphi_1(7,3,\omega_1,\omega_2) = -i126\omega_1 - j126\omega_2$	Filter
3	8	$\varphi_1(8,3,\omega_1,\omega_2) = -i144\omega_1 - j144\omega_2$	Filter
3	9	$\varphi_1(9,3,\omega_1,\omega_2) = -i172\omega_1 - j172\omega_2$	Filter
3	10	$\varphi_1(10,3,\omega_1,\omega_2) = -i180\omega_1 - j180\omega_2$	Filter
3	11	$\varphi_1(11,3,\omega_1,\omega_2) = -i198\omega_1 - j198\omega_2$	Filter
4	4	$\varphi_1(4,4,\omega_1,\omega_2) = -i96\omega_1 - j96\omega_2$	Filter
4	5	$\varphi_1(4,4,\omega_1,\omega_2) = -i120\omega_1 - j120\omega_2$	Filter
4	6	$\varphi_1(6,4,\omega_1,\omega_2) = -i144\omega_1 - j144\omega_2$	Filter
4	7	$\varphi_1(7,4,\omega_1,\omega_2) = -i168\omega_1 - j168\omega_2$	Filter
4	8	$\varphi_1(8,4,\omega_1,\omega_2) = -i192\omega_1 - j192\omega_2$	Filter
4	9	$\varphi_1(9,4,\omega_1,\omega_2) = --i216\omega_1 - 216\omega_2$	Filter
4	10	$\varphi_1(10,4,\omega_1,\omega_2) = -i240\omega_1 - j240\omega_2$	Filter
4	11	$\varphi_1(11,4,\omega_1,\omega_2) = -i264\omega_1 - j264\omega_2$	Filter

$$\varphi_1(M,K,\omega_1,\omega_2) = \varphi_1(7,3,\omega_1,\omega_2) = -i126\omega_1 - j126\omega_2 \qquad (68)$$

and it is shown in Figure 25. The amplitude response characteristics of the first novel class of 2D FIR filter in the range from $-\pi$ to π in absolute units and dB (3D plot) as well as the contour plot (2D) are given in Figures 26, 27, 28 and 29.

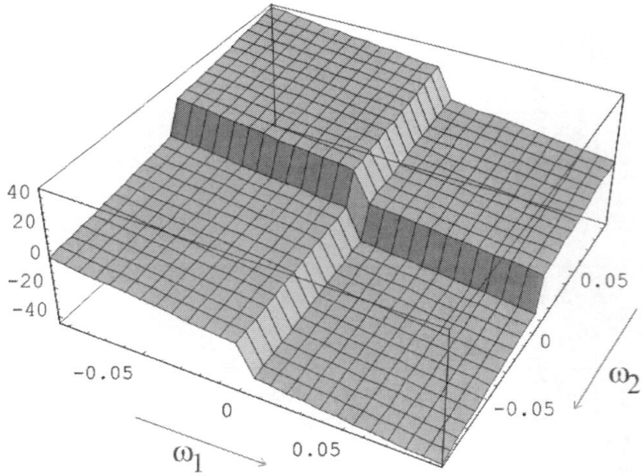

Figure 25. Phase response characteristics, $\varphi_1(7,3,\omega_1,\omega_2)$, of the first novel class of 2D FIR filter around the coordinate origin.

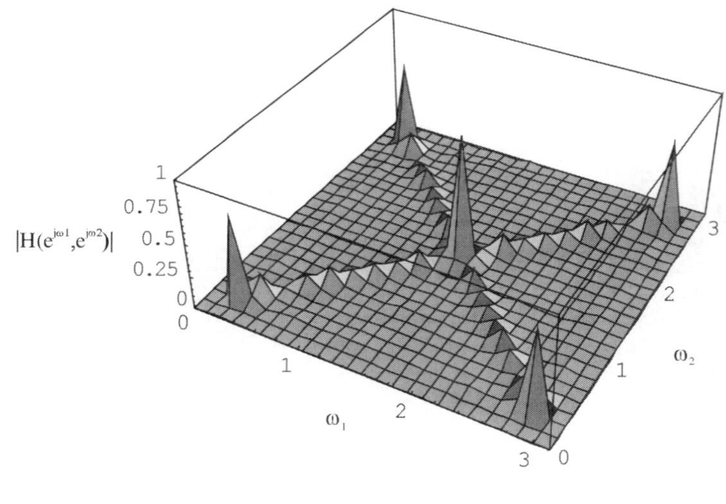

Figure 26. 3D plot of normalized amplitude characteristics of the first novel class of 2D FIR filter designed by Eq. (28) for M=7 and K=3 in the range $\omega_1 \in (0,\pi)$ and $\omega_2 \in (0,\pi)$.

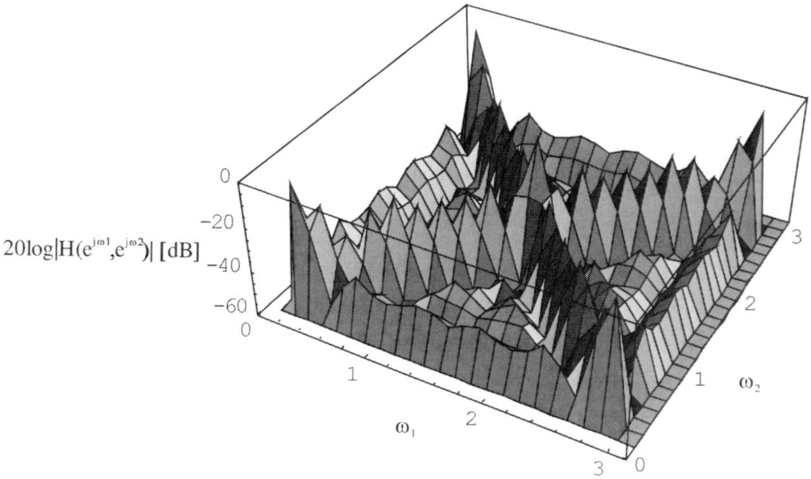

Figure 27. 3D plot of normalized magnitude characteristics in dB of the first novel class of 2D FIR filter generated by Eq. (29) for M=7 and K=3 in the range $\omega_1 \in (0, \pi)$ and $\omega_2 \in (0, \pi)$.

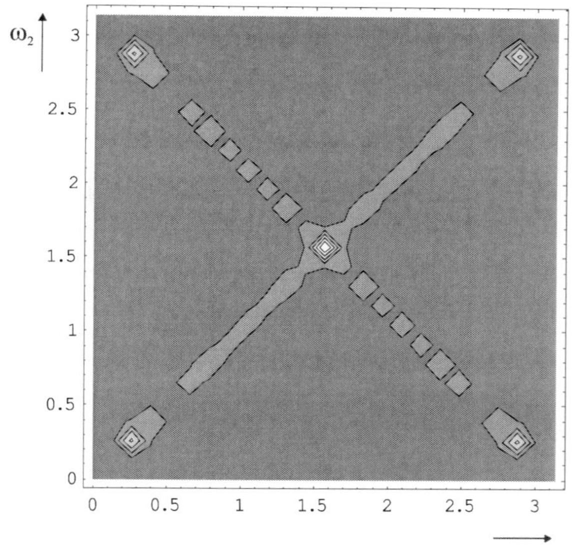

Figure 28. 2D contour plot of normalized magnitude characteristics of the first novel class of 2D FIR filter designed by Eq. (29) for M=7 and K=3 in the range $\omega_1 \in (0, \pi)$ and $\omega_2 \in (0, \pi)$.

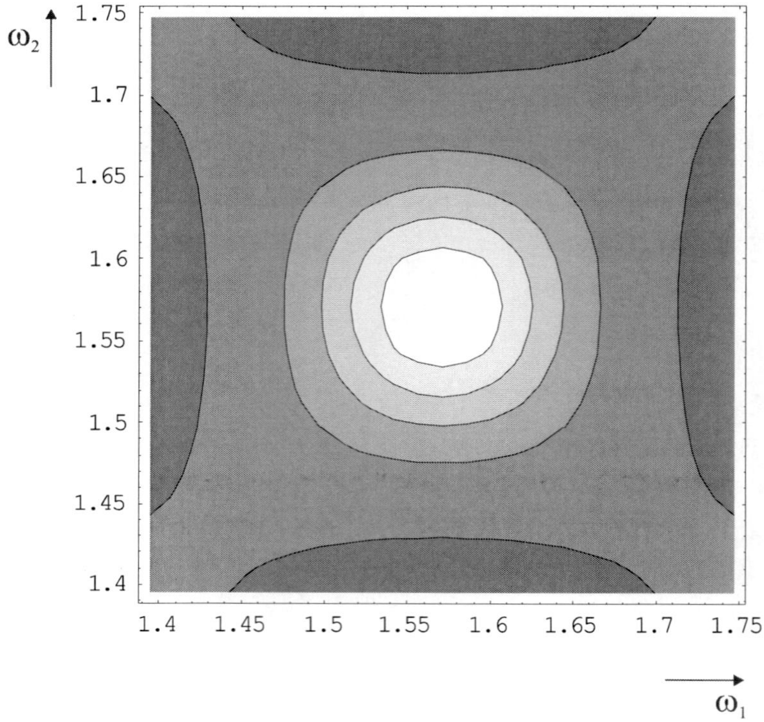

Figure 29. Zoomed Figure 28.

Example 2:

We consider the example of the proposed first novel class of 2D FIR filter for even M and odd K, M=8 and K=3. Phase response characteristics from Eq.(30) has the form:

$$\varphi_1(M, K, \omega_1, \omega_2) = \varphi_1(8,3,\omega_1,\omega_2) = -i144\omega_1 - j144\omega_2 \tag{69}$$

and it is shown in Figure 30.

The amplitude response characteristics of the first novel class of 2D FIR filter in the range from 0 to π in absolute units and dB (3D plot) as well as the contour plot (2D) are given in Figures 31, 32, 33 and 34 for even M and odd K, M=8 and K=3.

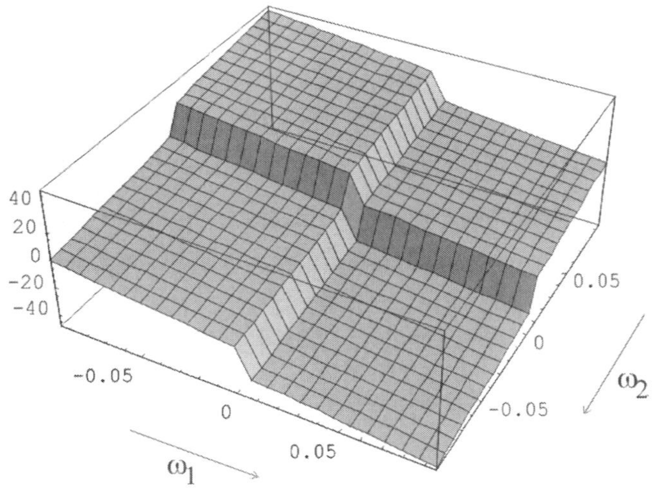

Figure 30. Phase response characteristics, $\varphi_1(8,3,\omega_1,\omega_2)$, of the first novel class of 2D FIR filter around coordinate origin.

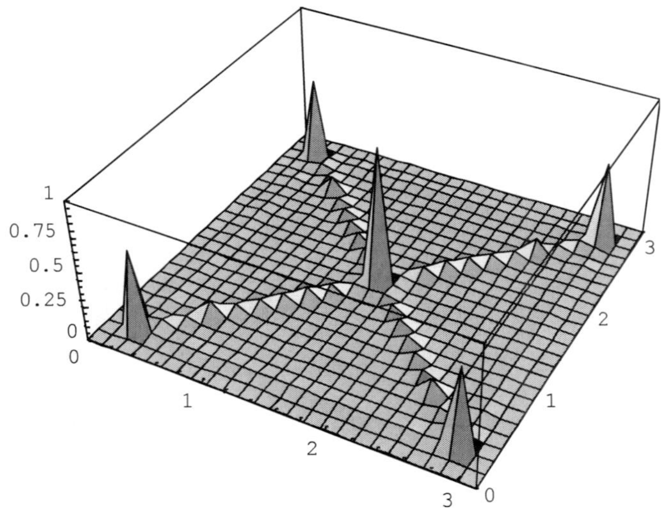

Figure 31. 3D plot of normalized amplitude characteristics of the first novel class of 2D FIR filter designed by Eq. (28) for M=8 and K=3 in the range $\omega_1 \in (0,\pi)$ and $\omega_2 \in (0,\pi)$.

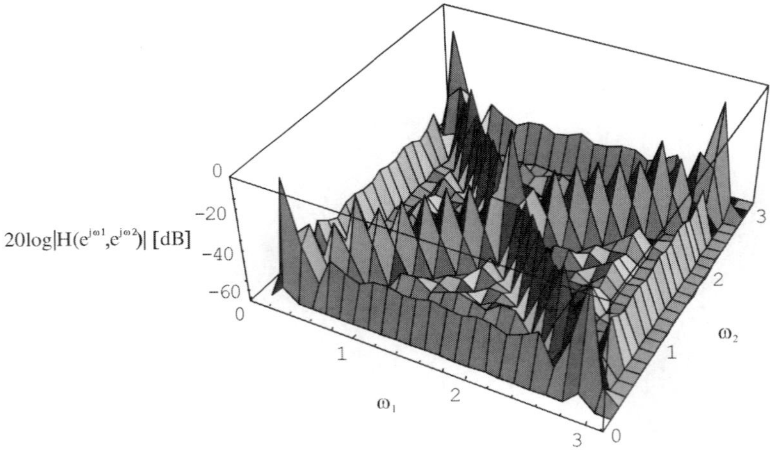

Figure 32. 3D plot of normalized magnitude characteristics in dB of the first novel class of 2D FIR filter designed by Eq. (29) for M=8 and K=3 in the range $\omega_1 \in (0, \pi)$ and $\omega_2 \in (0, \pi)$.

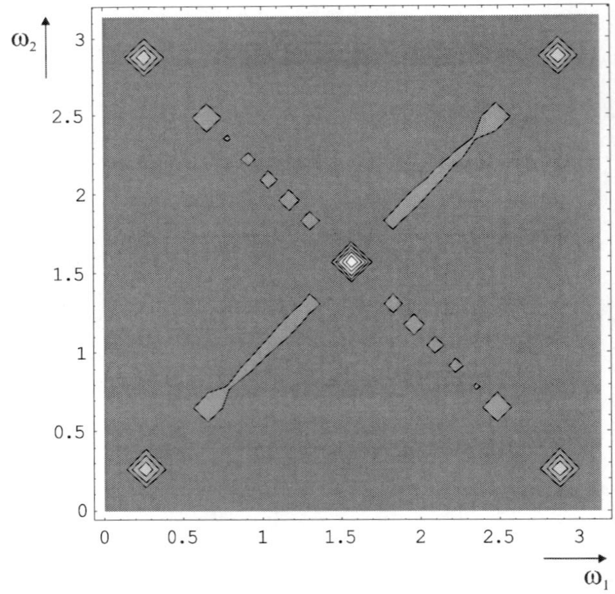

Figure 33. 2-D contour plot of normalized magnitude characteristics of the first novel class of 2D FIR filter designed by Eq. (29) for M=8 and K=3 in the range $\omega_1 \in (0, \pi)$ and $\omega_2 \in (0, \pi)$.

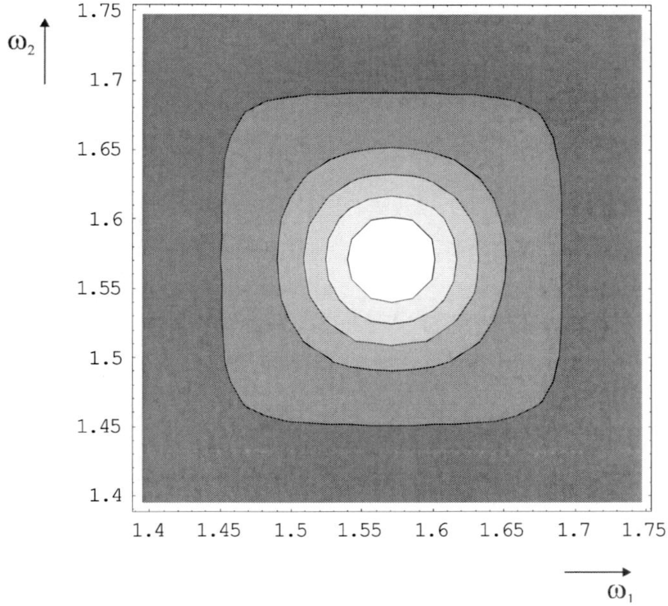

Figure 34. Zoomed Figure 33.

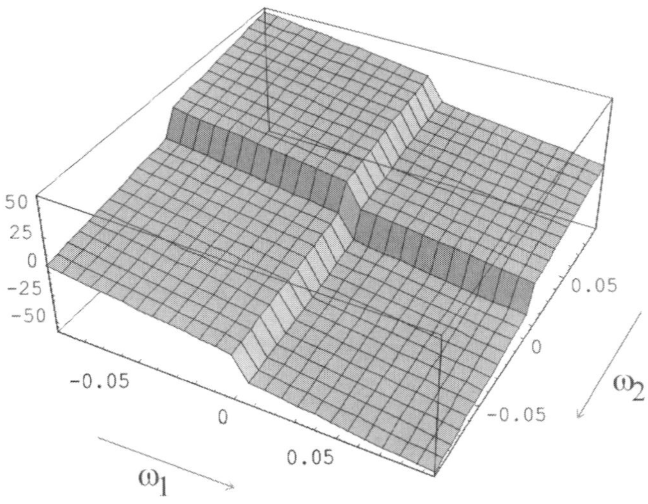

Figure 35. Phase response characteristics, $\varphi_1(11,3,\omega_1,\omega_2)$, of the first novel class of 2D FIR filter around coordinate origin.

Example 3:

We consider the example of the proposed first novel class of 2D FIR filter for odd M and K, M=11 and K=3. Phase response characteristics from Eq.(30) has the form:

$$\varphi_1(M, K, \omega_1, \omega_2) = \varphi_1(11,3,\omega_1,\omega_2) = -i198\,\omega_1 - j198\,\omega_2 \qquad (70)$$

and it is shown in Figure 35.

The amplitude response characteristics of this filter in the range from 0 to π in absolute units and dB (3D plot) as well as the contour plot (2D) are given in Figures 36, 37, 38 and 39 for odd M and K, M=11 and K=3.

Example 4:

We consider the example of the proposed first novel class of 2D FIR filter for even M and K, M=6 and K=4. Phase characteristics from Eq.(30) has a form

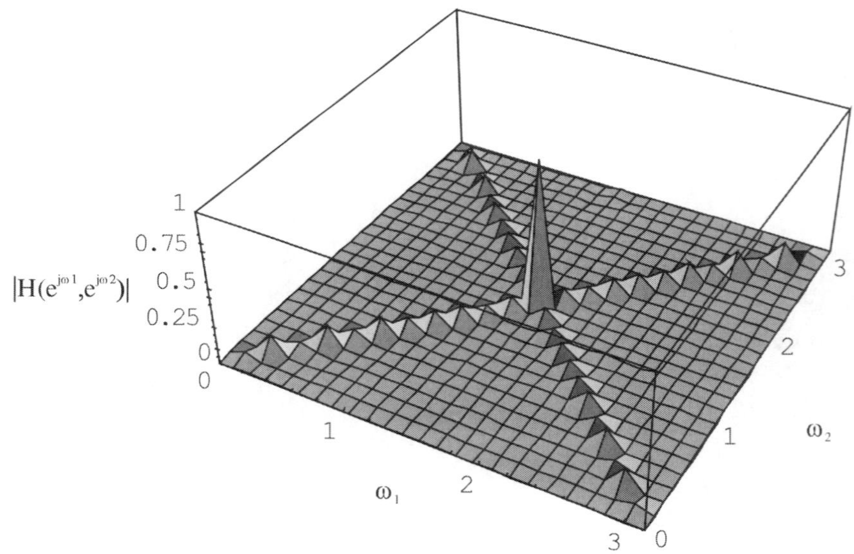

Figure 36. 3D plot of normalized amplitude characteristics of the first novel class of 2D FIR filter designed by Eq. (28) for M=11 and K=3 in the range $\omega_1 \in (0,\pi)$ and $\omega_2 \in (0,\pi)$.

$$\varphi_1(M,K,e^{j\omega_1},e^{j\omega_2}) = \varphi_1(6,4,\omega_1,\omega_2) = -i144\omega_1 - j144\omega_2 \quad (70)$$

and it is shown in Figure 40.

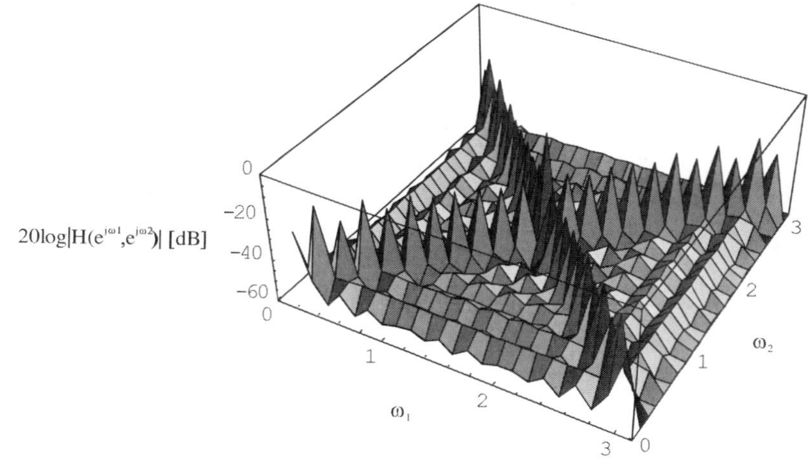

Figure 37. 3D plot of normalized magnitude characteristics in dB of the first novel class of 2D FIR filter designed by Eq. (29) for M=11 and K=3 in the range $\omega_1 \in (0,\pi)$ and $\omega_2 \in (0,\pi)$.

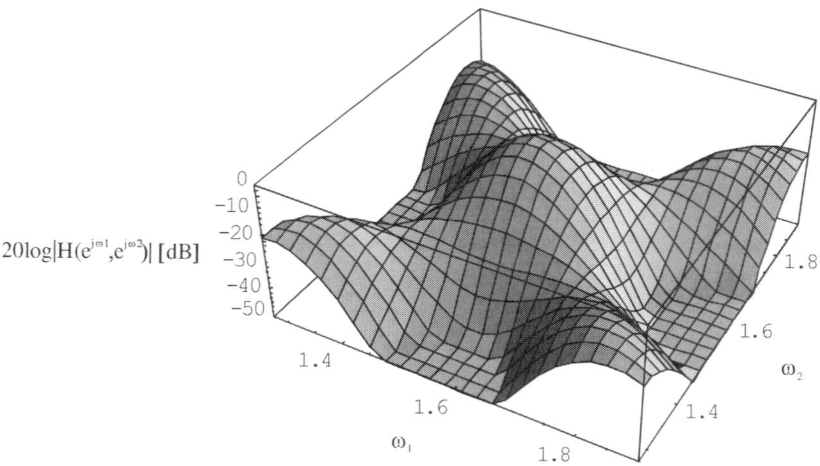

Figure 38. Zoomed Figure 37.

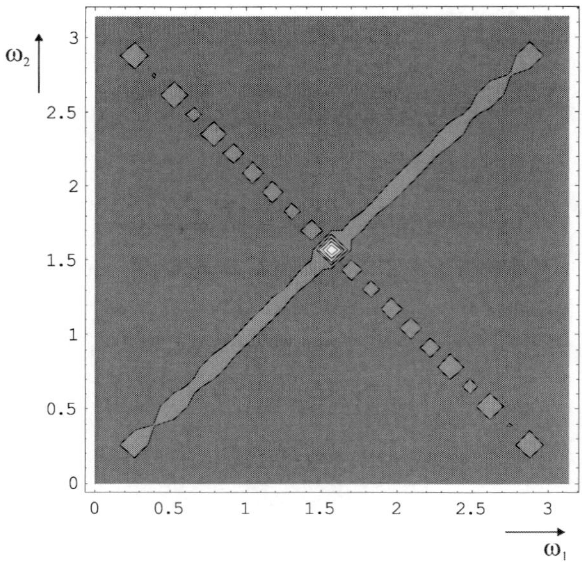

Figure 39. 2D contour plot of normalized magnitude characteristics of the first novel class of 2D FIR filter designed by Eq. (29) for M=11 and K=3 in the range $\omega_1 \in (0, \pi)$ and $\omega_2 \in (0, \pi)$.

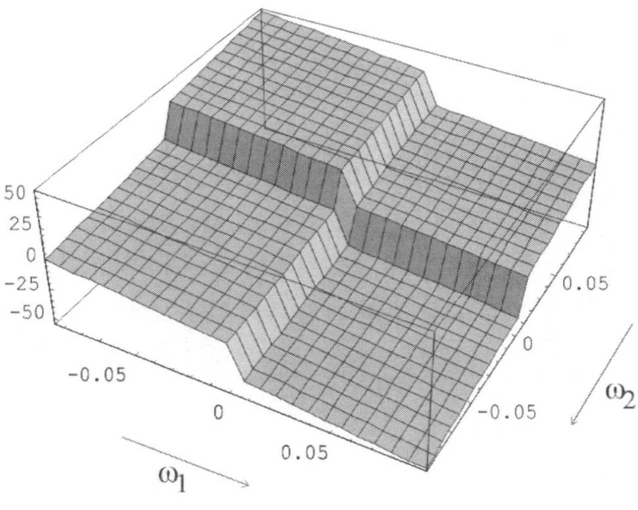

Figure 40. Phase characteristics around coordinate origin, $\varphi_1(6{,}4,\omega_1,\omega_2)$, of the first novel class of 2D FIR filter around coordinate origin.

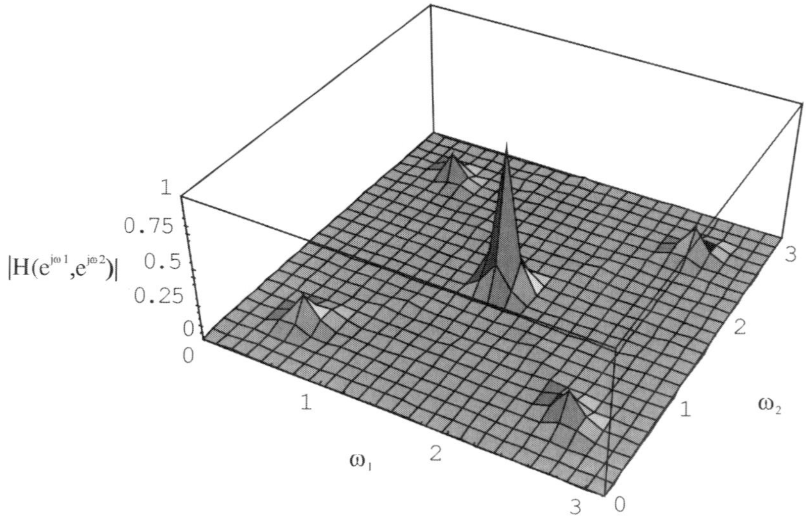

Figure 41. 3D plot of normalized amplitude characteristics of the first novel class of 2D FIR filter designed by Eq. (28) for *M*=6 and *K*=4 in the range $\omega_1 \in (0,\pi)$ and $\omega_2 \in (0,\pi)$.

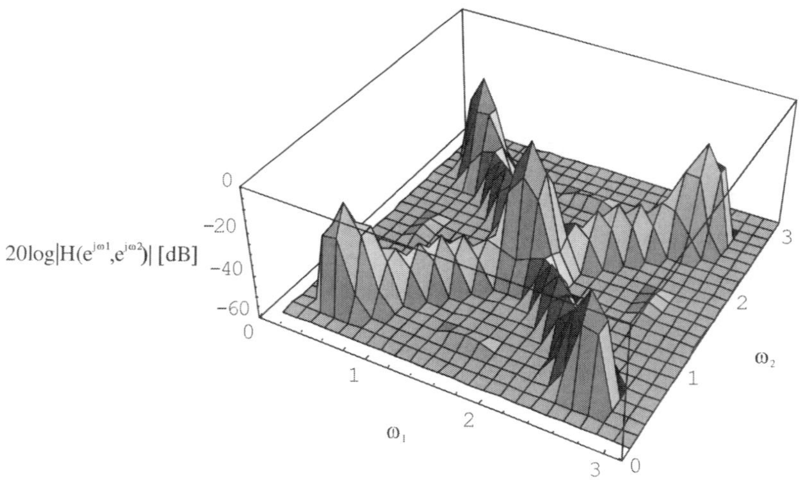

Figure 42. 3D plot of normalized magnitude characteristics in dB of the first novel class of 2D FIR filter designed by Eq. (29) for *M*=6 and *K*=4 in the range $\omega_1 \in (0,\pi)$ and $\omega_2 \in (0,\pi)$.

The amplitude response characteristics of this filter in the range from 0 to π in absolute units and dB (3D plot) as well as the contour plot (2D plot) are given in Figures 41, 42, 43, 44 and 45 for even *M* and *K*, *M*=6 and *K*=4.

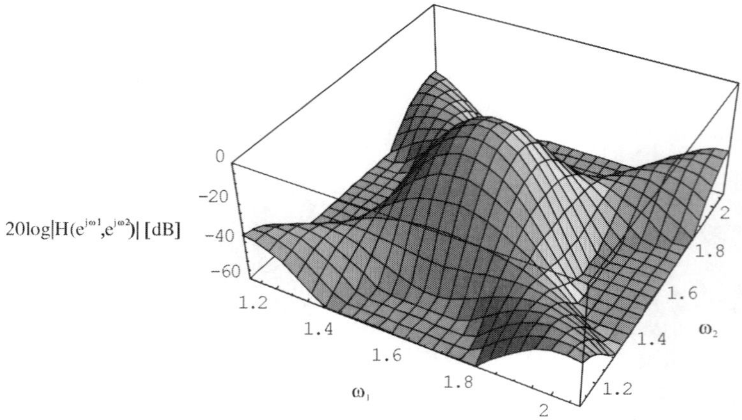

Figure 43. Zoomed Figure 42.

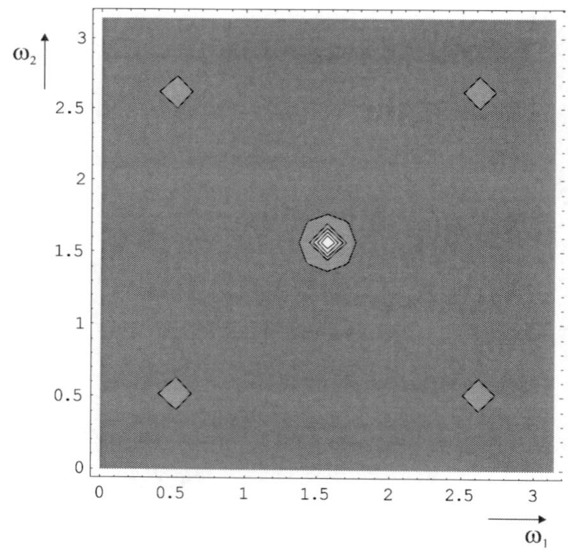

Figure 44. 2D contour plot of normalized magnitude characteristics of the first novel class of 2D FIR filter designed by Eq. (29) for *M*=6 and *K*=4 in the range $\omega_1 \in (0,\pi)$ and $\omega_2 \in (0,\pi)$.

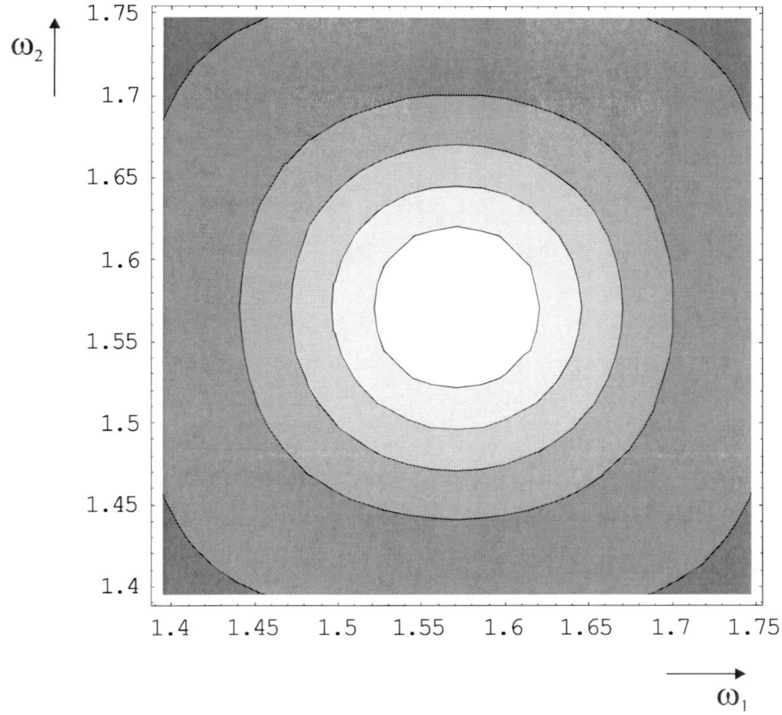

Figure 45. Zoomed Figure 44.

Table 3. Normalized value of the pass-band surface of the first novel class of 2D FIR filter, for the some values of *M* and *K*, for the attenuation value of 0.28 dB at the cut-off frequency of the pass-band

Attenuation value at the cut-off frequency of the pass-band	M	K	Normalized value of the pass-band surface
0.28 dB	5	3	0.0000983200901484082
0.28 dB	6	3	0.00006914985385806480
0.28 dB	9	3	0.00003169977211427059

For the some values of *M* and *K*, normalized value of the pass-band surface of the first novel class of 2D FIR filter is given in Table 3 for the attenuation value of 0.28 dB at the cut-off frequency of the pass-band. Normalized surface in % of the stop-band of the first novel class of 2D FIR filter, for the some values of *M* and *K*, is given in Table 4 for the attenuation values in the stop-band of 60 dB, 80 dB and 100 dB.

7. Examples of the Second Novel Class of Linear Phase and Non-Recursive 2D FIR Filter Functions

Examples of the proposed second novel class of 2D FIR filter functions obtained for particular values of free real integer parameters *K* and *M* from Eq. (32), applying Eq. (26) are given next in the chapter.

Example 5:

We consider the example of proposed second novel class of 2D FIR filter for odd *M* and *K*, *M*=5 and *K*=3.

The amplitude response characteristics of this filter in the range from 0 to π in absolute units and dB (3D plot) as well as the contour plot (2D plot) are given in Figures 46, 47, 48, 49 50 and 51 for *M*=5 and *K*=3.

Table 4. Normalized surface in % of the stop-band of the first novel class of 2D FIR filter, for the some values of *M* and *K*, for the attenuation values in the stop-band of 60 dB, 80 dB and 100 dB

Attenuation value in the stop-band	M	K	Normalized surface in % of the stop-band
60 dB	5	3	77.2235208181154%
80 dB	5	3	69.3769174579985%
100 dB	5	3	56.4864864864865%
60 dB	6	3	81.5434623813002%
80 dB	6	3	75.9495982468955%
100 dB	6	3	66.1723886048210%
60 dB	9	3	86.1796932067202%
80 dB	9	3	85.2249817384953%
100 dB	9	3	72.8137326515705%

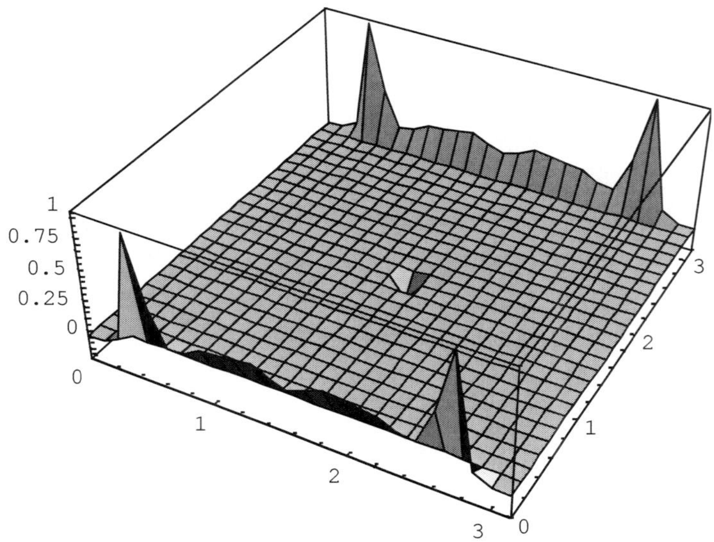

Figure 46. Normalized amplitude response characteristics of the second novel class of 2D FIR filter, designed by Eq.(47), for M=5 and K=3 in the range $\omega_1 \in (0,\pi)$ and $\omega_2 \in (0,\pi)$.

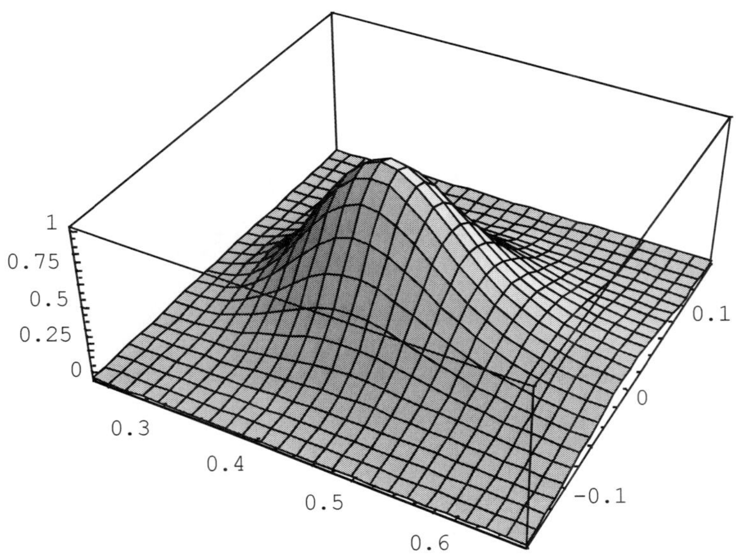

Figure 47. Zoomed Figure 46.

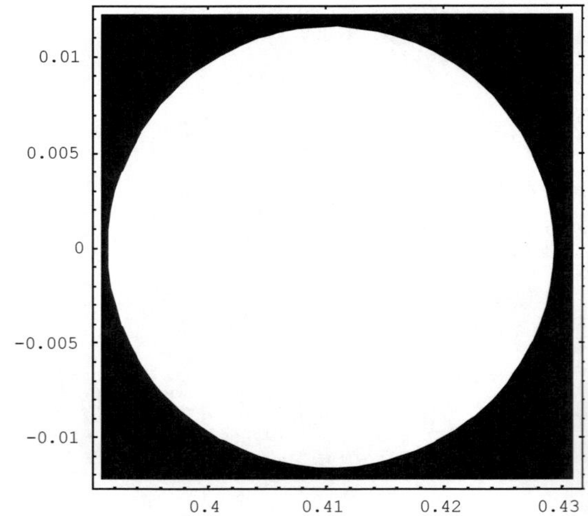

Figure 48. The shape of pass-band surface of the second novel class of 2D FIR filter, designed by Eq.(47), for M=5 and K=3, for the attenuation value of 0.28 dB at the cut-off frequency of pass-band.

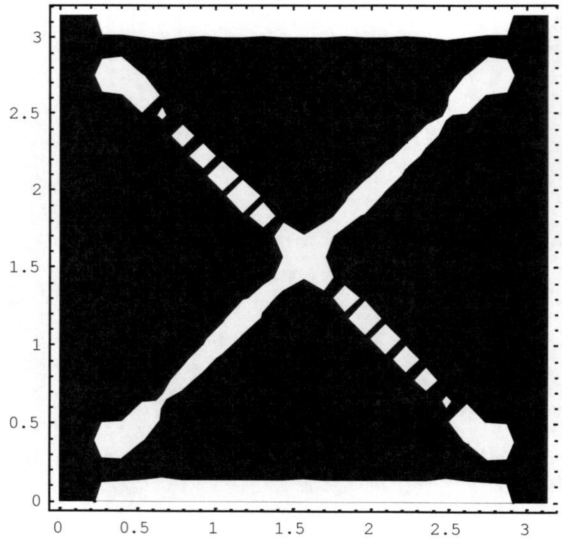

Figure 49. The shape of stop-band surface of the second novel class of 2D FIR filter, designed by Eq.(48), for M=5 and K=3 for the attenuation value of 80 dB in the stop-band in the range $\omega_1 \in (0, \pi)$ and $\omega_2 \in (0, \pi)$.

Figure 50. The shape of stop-band surface of the second novel class of 2D FIR filter, designed by Eq.(48), for *M*=5 and *K*=3 for the attenuation value of 160 dB in the stop-band in the range $\omega_1 \in (0, \pi)$ and $\omega_2 \in (0, \pi)$.

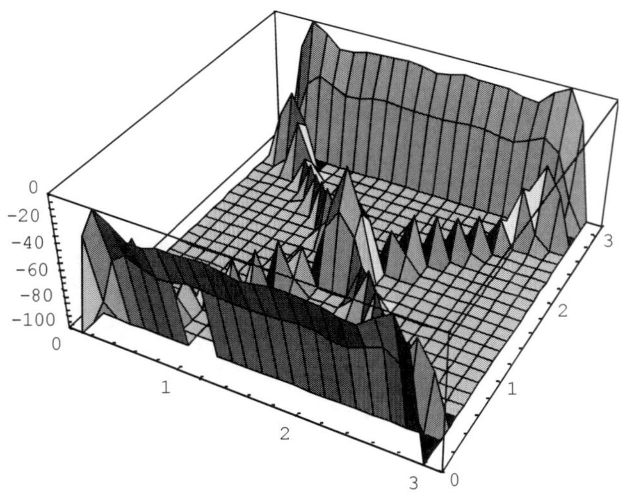

Figure 51. Normalized magnitude response characteristics in dB of the second novel class of 2D FIR filter, designed by Eq.(48), for *M*=5 and *K*=3 in the range $\omega_1 \in (0, \pi)$ and $\omega_2 \in (0, \pi)$.

Example 6:

We consider the example of proposed second novel class of 2D FIR filter for even M and odd K, M=6 and K=3.

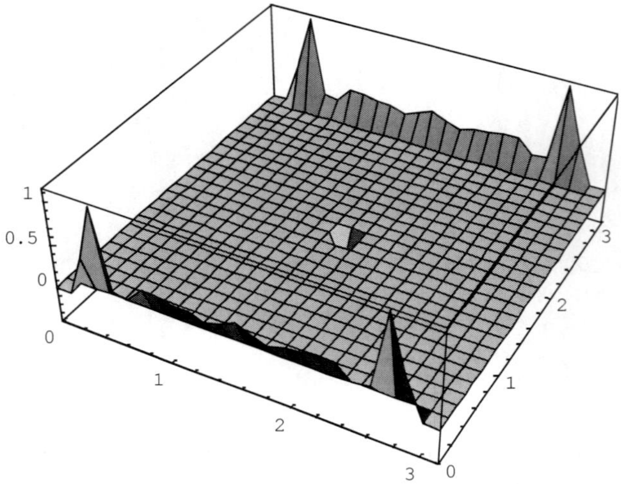

Figure 52. Normalized amplitude response characteristics of the second novel class of 2D FIR filter, designed by Eq.(47), for M=6 and K=3 in the range $\omega_1 \in (0,\pi)$ and $\omega_2 \in (0,\pi)$.

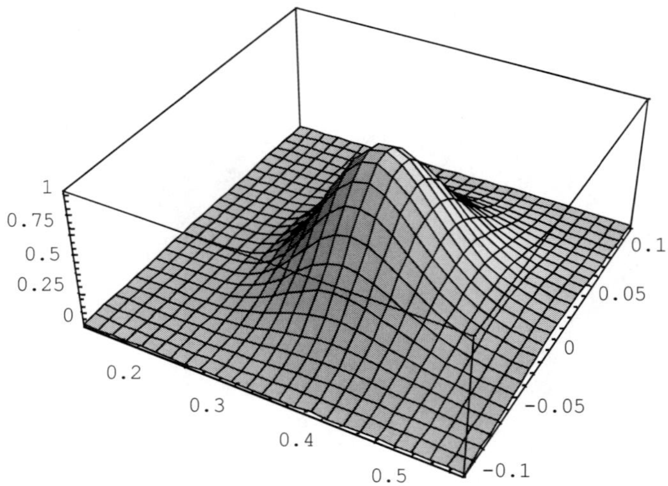

Figure 53. Zoomed Figure 52.

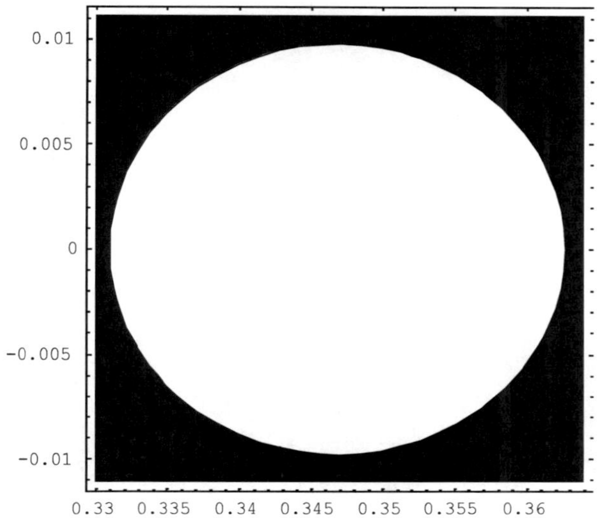

Figure 54. The shape of pass-band surface of the second novel class of 2D FIR filter, designed by Eq.(48), for M=6 and K=3, for the attenuation value of 0.28 dB at the cut-off frequency of pass-band.

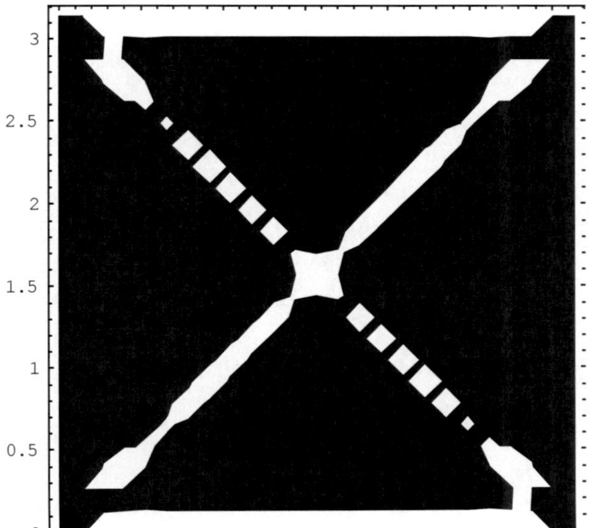

Figure 55. The shape of stop-band surface of the second novel class of 2D FIR filter, designed by Eq.(48), for M=6 and K=3 for the attenuation value of 80 dB in the stop-band in the range $\omega_1 \in (0, \pi)$ and $\omega_2 \in (0, \pi)$.

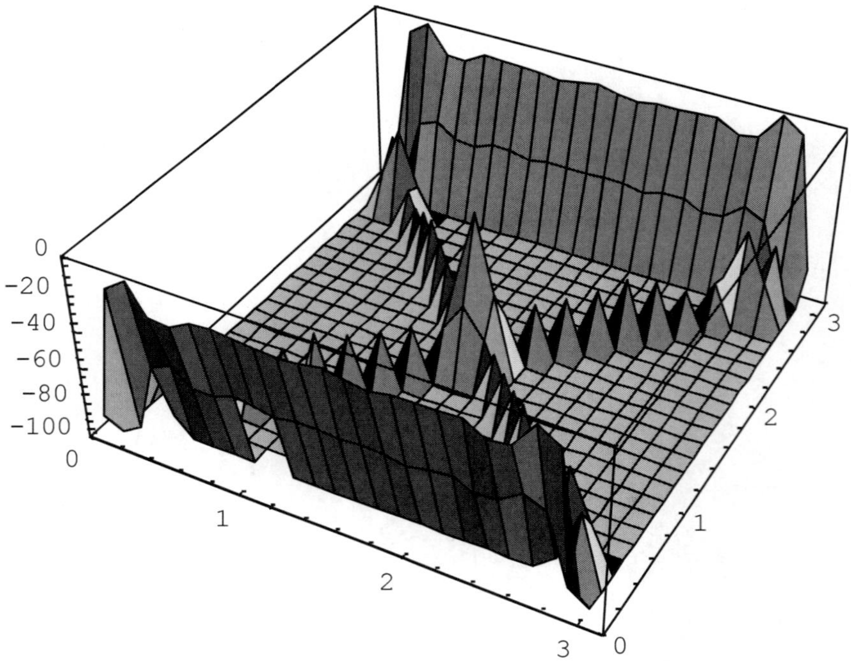

Figure 56. Normalized magnitude response characteristics in dB of the second novel class of 2D FIR filter, designed by Eq.(48), for M=6 and K=3 in the range $\omega_1 \in (0, \pi)$ and $\omega_2 \in (0, \pi)$.

The amplitude response characteristics of this filter in the range from 0 to π in absolute units and dB (3D plot) as well as the contour plot (2D plot) are given in Figures 52, 53, 54, 55 and 56 for M=6 and K=3.

Example 7:

We consider the example of proposed second novel class of 2D FIR filter for odd M and K, M=9 and K=3.

The amplitude response characteristics of this filter in the range from 0 to π in absolute units and dB (3D plot) as well as the contour plot (2D plot) are given in Figure 57, Figure 58, Figure 59, Figure 60, Figure 61 and Figure 62 for M=9 and K=3.

For some values of M and K, normalized value of the pass-band surface of the second novel class of 2D FIR filter is given in Table 5 for the attenuation value of 0.28 dB at the cut-off frequency of the pass-band.

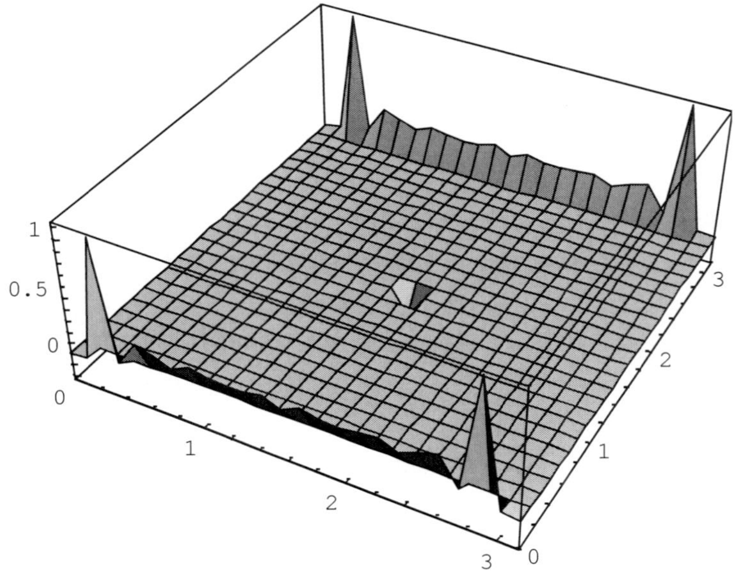

Figure 57. Normalized amplitude response characteristics of the second novel class of 2D FIR filter, designed by Eq.(47), for *M*=9 and *K*=3 in the range $\omega_1 \in (0,\pi)$ and $\omega_2 \in (0,\pi)$.

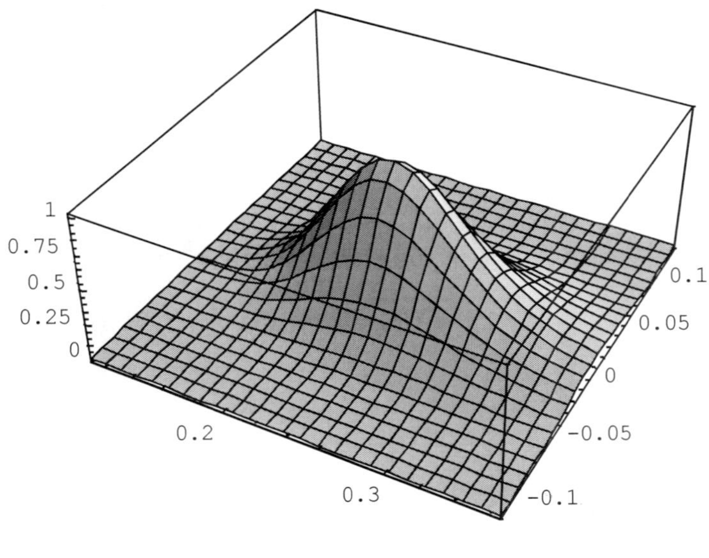

Figure 58. Zoomed Figure 57.

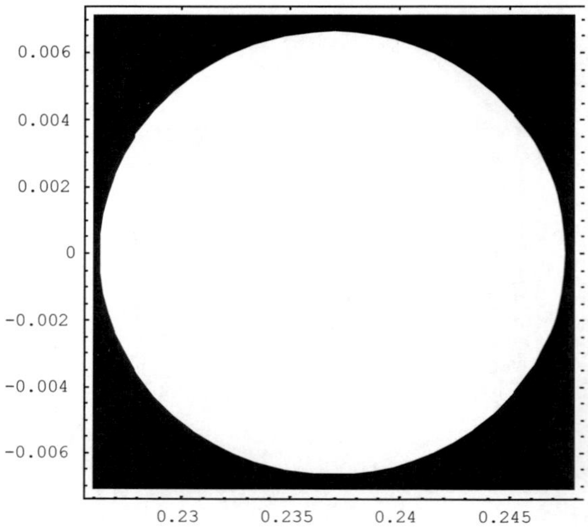

Figure 59. The shape of pass-band surface of the second novel class of 2D FIR filter, designed by Eq.(48), for M=9 and K=3, for the attenuation value of 0.28 dB at the cut-off frequency of pass-band.

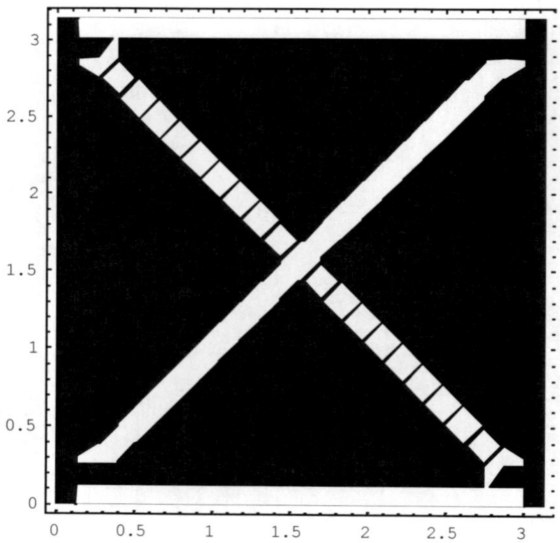

Figure 60. The shape of stop-band surface of the second novel class of 2D FIR filter, designed by Eq.(48), for M=9 and K=3 for the attenuation value of 100 dB in the stop-band in the range $\omega_1 \in (0, \pi)$ and $\omega_2 \in (0, \pi)$.

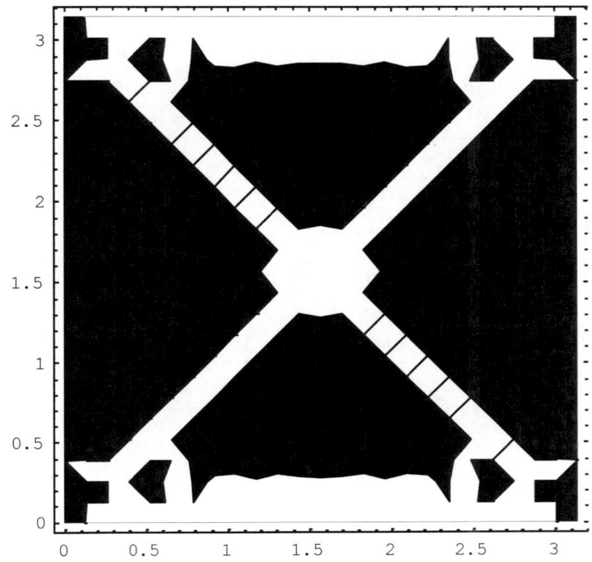

Figure 61. The shape of stop-band surface of the second novel class of 2D FIR filter, designed by Eq.(48), for M=9 and K=3 for the attenuation value of 160 dB in the stop-band in the range $\omega_1 \in (0, \pi)$ and $\omega_2 \in (0, \pi)$.

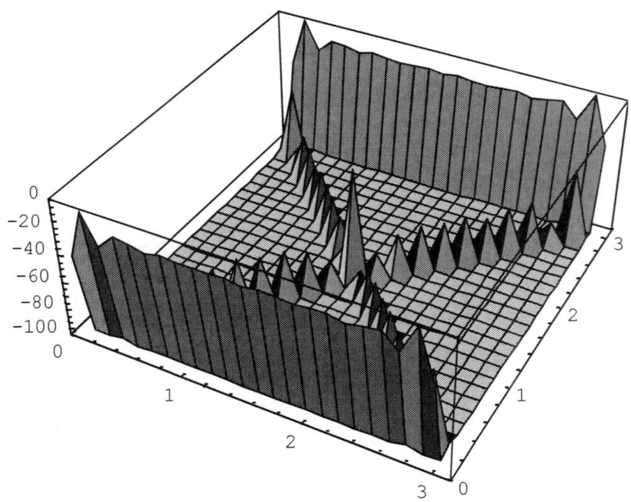

Figure 62. Normalized magnitude response characteristics in dB of the second novel class of 2D FIR filter, designed by Eq.(48), for M=9 and K=3 in the range $\omega_1 \in (0, \pi)$ and $\omega_2 \in (0, \pi)$.

Table 5. Normalized value of the pass-band surface of the second novel class of 2D FIR filter, for some values of *M* and *K*, for the attenuation value of 0.28 dB at the cut-off frequency of the pass-band

Attenuation value at the cut-off frequency of the pass-band	M	K	Normalized value of the pass-band surface
0.28 dB	5	3	0.00006954471561068209
0.28 dB	6	3	0.00004888956325294961
0.28 dB	9	3	0.00002240090522062185

Table 6. Normalized surface in % of the stop-band of the second novel class of 2D FIR filter, for some values of *M* and *K*, for the attenuation values in the stop-band of 60 dB, 80 dB and 100 dB

Attenuation value in the stop-band	M	K	Normalized surface in % of the stop-band
60 dB	5	3	90.2921840759679%
80 dB	5	3	82.7552958363769%
100 dB	5	3	78.4360847333820%
60 dB	6	3	91.1285609934259%
80 dB	6	3	83.9780861943024%
100 dB	6	3	78.5624543462381%
60 dB	9	3	91.9459459459459%
80 dB	9	3	85.5792549306063%
100 dB	9	3	80.1453615777940%

Normalized surface in % of the stop-band of the second novel class of 2D FIR filter, for some values of *M* and *K*, is given in Table 6 for the attenuation values in the stop-band of 60 dB, 80 dB and 100 dB.

8. Examples of the Third Novel Class of Linear Phase and Non-Recursive 2D FIR Filter Functions

Examples of the proposed third novel class of 2D FIR filter functions obtained for particular values of free real integer parameters K and M from Eq. (51), applying Eq. (26) are given next in the chapter.

Example 8:

We consider the example of proposed third novel class of 2D FIR filter for odd M and K, M=5 and K=3.

The amplitude response characteristics of this filter in the range from 0 to π in absolute units and dB (3D plot) as well as the contour plot (2D plot) are given in Figures 63, 64, 65, 66 and 67 for M=5 and K=3.

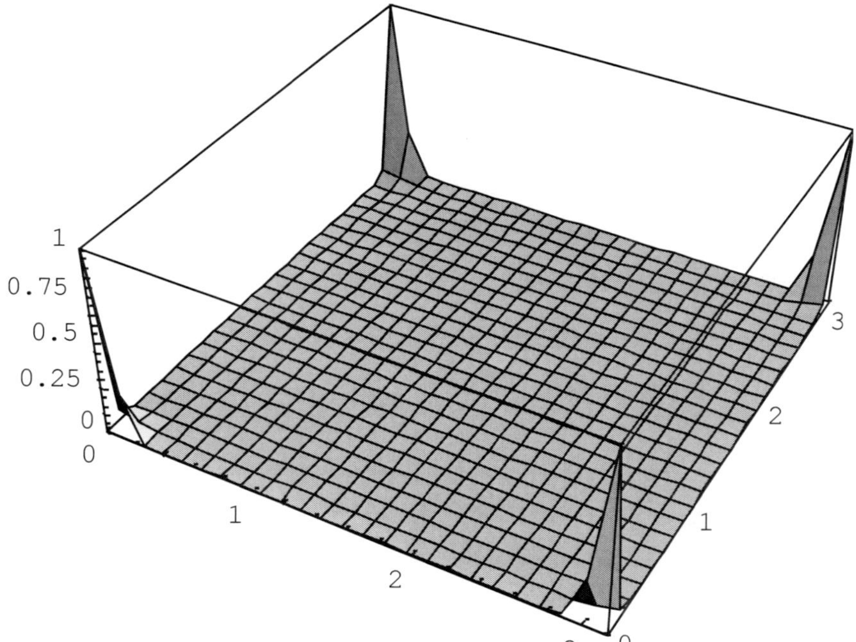

Figure 63. Normalized amplitude response characteristics of the third novel class of 2D FIR filter, designed by Eq.(65), for M=5 and K=3 in the range $\omega_1 \in (0, \pi)$ and $\omega_2 \in (0, \pi)$.

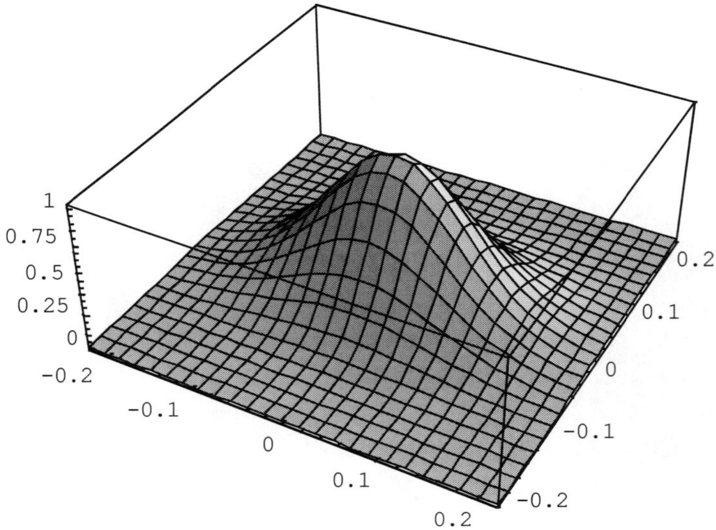

Figure 64. Zoomed Figure 63.

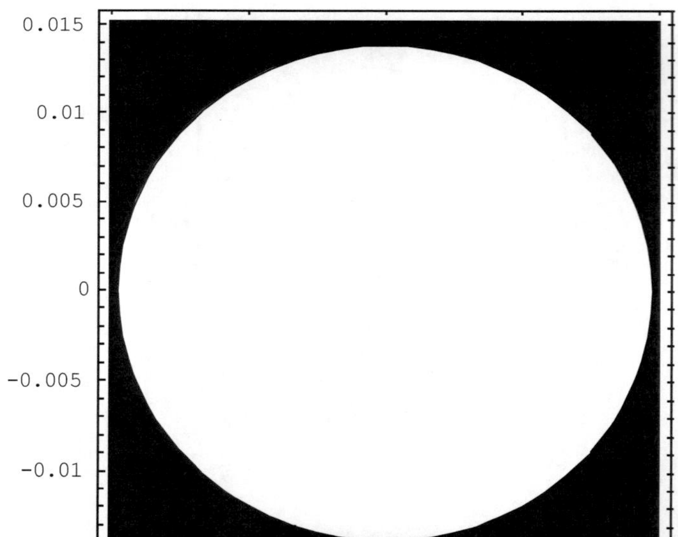

Figure 65. The shape of pass-band surface of the third novel class of 2D FIR filter, designed by Eq.(66), for M=5 and K=3, for the attenuation value of 0.28 dB at the cut-off frequency of pass-band.

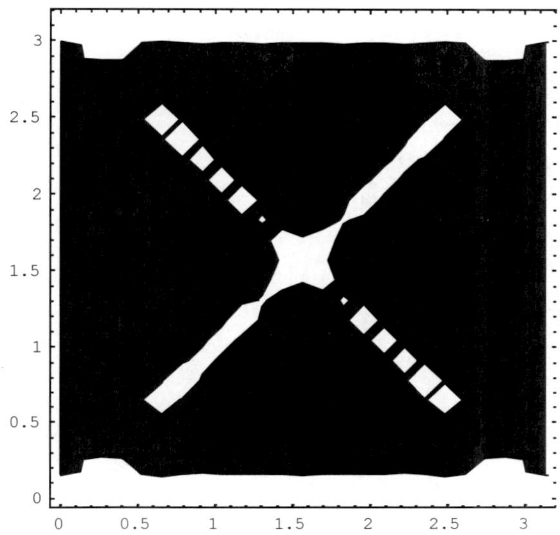

Figure 66. The shape of stop-band surface of the third novel class of 2D FIR filter, designed by Eq.(66), for *M*=5 and *K*=3 for the attenuation value of 100 dB in the stop-band in the range $\omega_1 \in (0, \pi)$ and $\omega_2 \in (0, \pi)$.

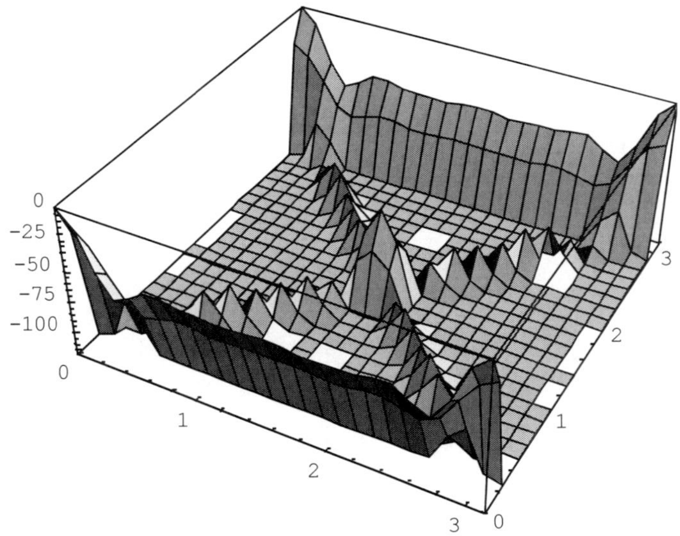

Figure 67. Normalized magnitude response characteristics in dB of the third novel class of 2D FIR filter, designed by Eq.(66), for M=5 and K=3 in the range $\omega_1 \in (0, \pi)$ and $\omega_2 \in (0, \pi)$.

Example 9:

We consider the example of proposed third novel class of 2D FIR filter for even M and odd K, M=6 and K=3.

The amplitude response characteristics of this filter in the range from 0 to π in absolute units and dB (3D plot) as well as the contour plot (2D plot) are given in Figures 68, 69, 70, 71 and 72 for M=6 and K=3.

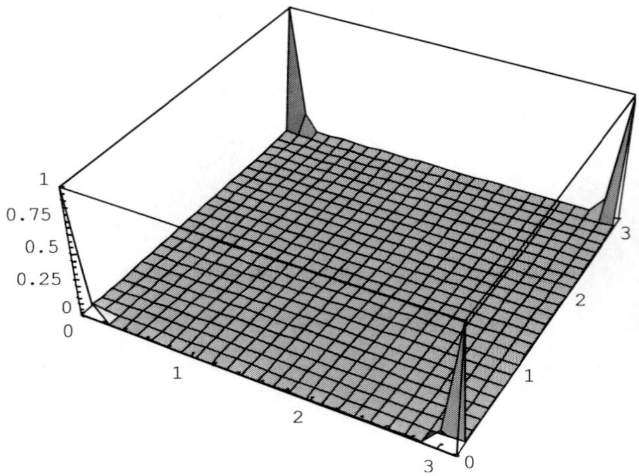

Figure 68. Normalized amplitude response characteristics of the third novel class of 2D FIR filter, designed by Eq.(65), for M=6 and K=3 in the range $\omega_1 \in (0, \pi)$ and $\omega_2 \in (0, \pi)$.

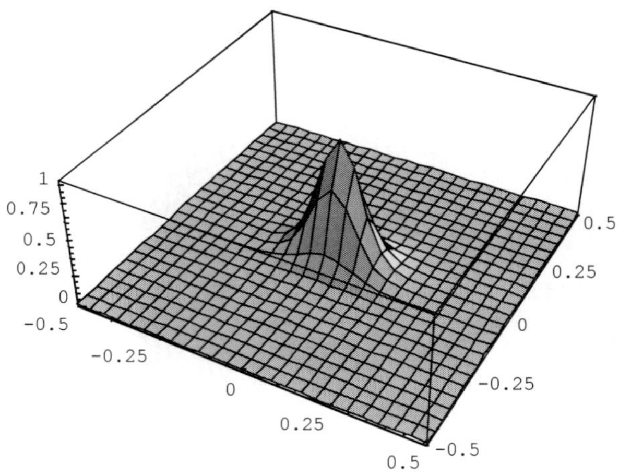

Figure 69. Zoomed Figure 68.

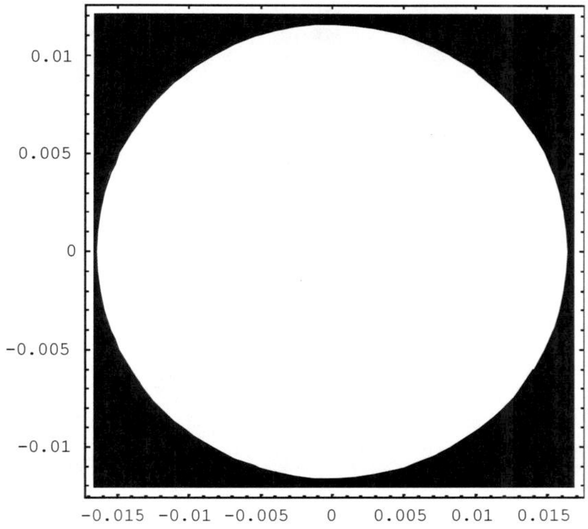

Figure 70. The shape of pass-band surface of the third novel class of 2D FIR filter, designed by Eq.(66), for M=6 and K=3, for the attenuation value of 0.28 dB at the cut-off frequency of pass-band.

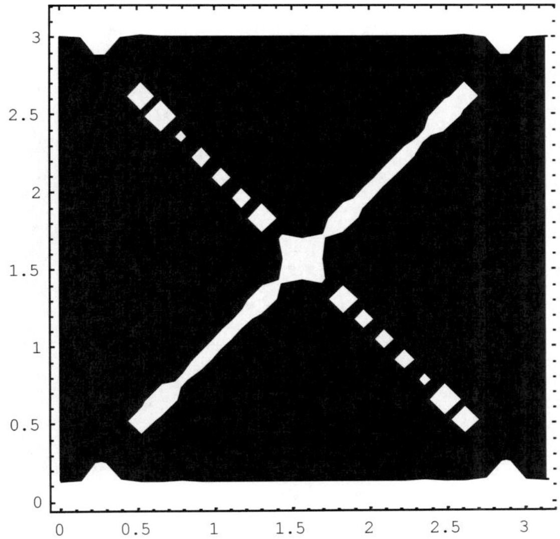

Figure 71. The shape of stop-band surface of the third novel class of 2D FIR filter, designed by Eq.(66), for M=6 and K=3 for the attenuation value of 100 dB in the stop-band in the range $\omega_1 \in (0, \pi)$ and $\omega_2 \in (0, \pi)$.

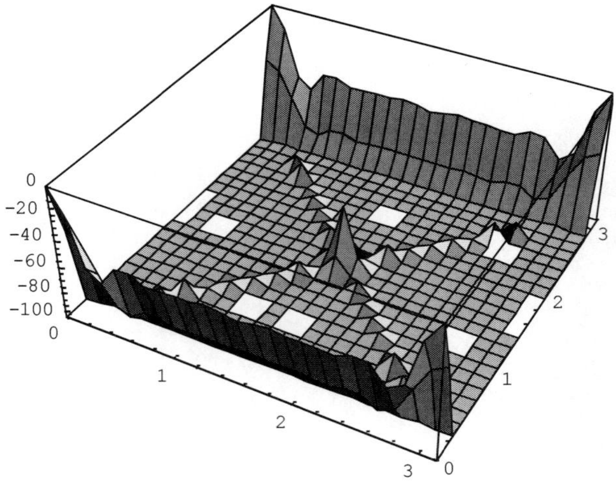

Figure 72. Normalized magnitude response characteristics in dB of the third novel class of 2D FIR filter, designed by Eq.(66), for M=6 and K=3 in the range $\omega_1 \in (0, \pi)$ and $\omega_2 \in (0, \pi)$.

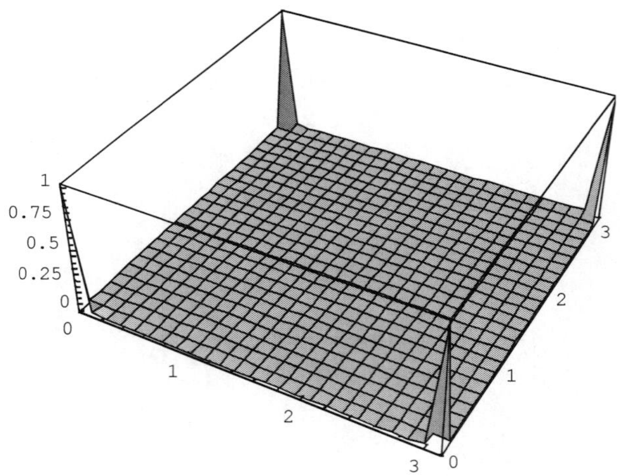

Figure 73. Normalized amplitude response characteristics of the third novel class of 2D FIR filter, designed by Eq.(65), for M=9 and K=3 in the range $\omega_1 \in (0, \pi)$ and $\omega_2 \in (0, \pi)$.

Example 10:

We consider the example of proposed third novel class of 2D FIR filter for odd M and K, M=9 and K=3.

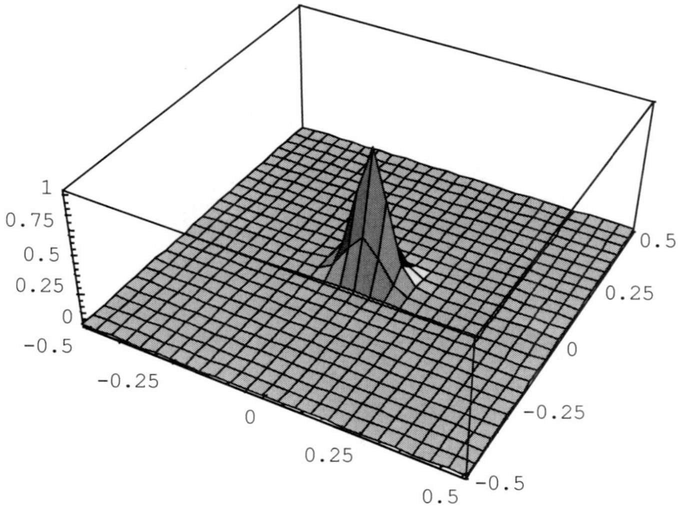

Figure 74. Zoomed Figure 73.

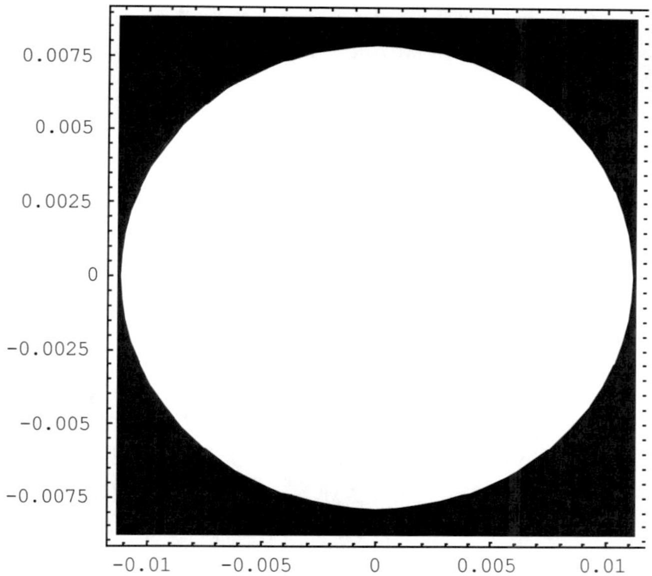

Figure 75. The shape of pass-band surface of the third novel class of 2D FIR filter, designed by Eq.(66), for M=9 and K=3, for the attenuation value of 0.28 dB at the cut-off frequency of pass-band.

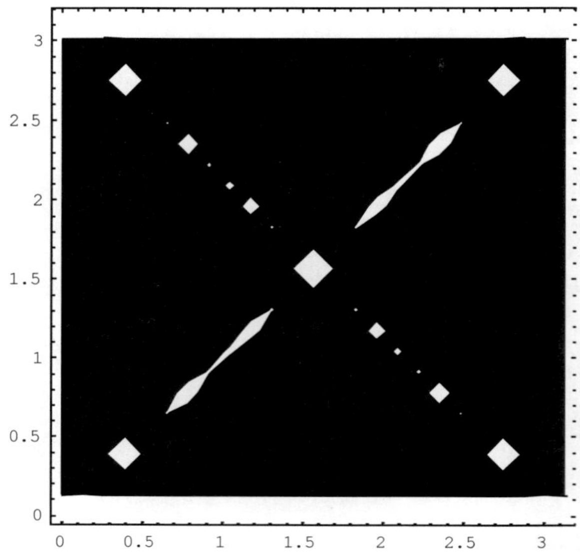

Figure 76. The shape of stop-band surface of the third novel class of 2D FIR filter, designed by Eq.(66), for M=9 and K=3 for the attenuation value of 100 dB in the stop-band in the range $\omega_1 \in (0,\pi)$ and $\omega_2 \in (0,\pi)$.

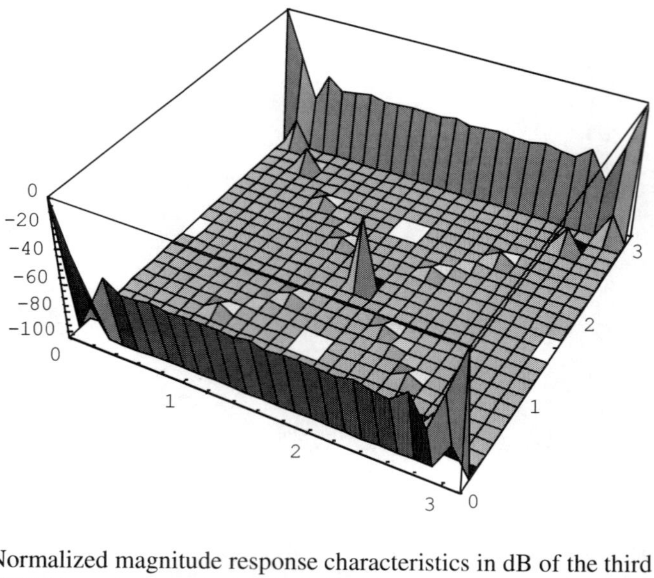

Figure 77. Normalized magnitude response characteristics in dB of the third novel class of 2D FIR filter, designed by Eq.(66), for M=9 and K=3 in the range $\omega_1 \in (0,\pi)$ and $\omega_2 \in (0,\pi)$.

The amplitude response characteristics of this filter in the range from 0 to π in absolute units and dB (3D plot) as well as the contour plot (2D plot) are given in Figures 73, 74, 75, 76 and 77 for M=9 and K=3.

For some values of M and K, normalized value of the pass-band surface of the third novel class of 2D FIR filter is given in Table 7 for the attenuation value of 0.28 dB at the cut-off frequency of the pass-band.

Normalized surface in % of the stop-band of the third novel class of 2D FIR filter, for some values of M and K, is given in Table 8 for the attenuation values in the stop-band of 60 dB, 80 dB and 100 dB.

Table 7. Normalized value of the pass-band surface of the third novel class of 2D FIR filter, for some values of *M* and *K*, for the attenuation value of 0.28 dB at the cut-off frequency of the pass-band

Attenuation value at the cut-off frequency of the pass-band	M	K	Normalized value of the pass-band surface
0.28 dB	5	3	0.000094501246630152O3
0.28 dB	6	3	0.0000598856616342422 5
0.28 dB	9	3	0.0000277966154419039 2

Table 8. Normalized surface in % of the stop-band of the third novel class of 2D FIR filter, for some values of *M* and *K*, for the attenuation values in the stop-band of 60 dB, 80 dB and 100 dB

Attenuation value in the stop-band	M	K	Normalized surface in % of the stop-band
60 dB	5	3	94.2227903579255%
80 dB	5	3	89.2140248356465%
100 dB	5	3	82.4667640613587%
60 dB	6	3	95.1249086924763%
80 dB	6	3	91.3250547845142%
100 dB	6	3	85.0146092037984%
60 dB	9	3	96.0511322132944%
80 dB	9	3	92.2783053323594%
100 dB	9	3	89.1986851716581%

Conclusion

In this chapter, three novel classes of non-recursive multiplierless linear phase 2D FIR filter functions with ultra selective pass-band spike are described. Proposed FIR filter functions are of very low complexity as they do not require multipliers for their realization.

Future applications in the THz frequency range and new modern wireless communications will require filter functions with significant selectivity, good suppression of unwanted Gibss phenomena and very narrow pass-band. We expect numerous 3D wireless applications in future.

References

[1] Abramowitz, M., and Stegun, I. (1964), Handbook on Mathematical Function, National Bureau of Standards, Applied Mathematics Series, USA.

[2] Szegö, G. (1967), Orthogonal Polynomials, 3rd Edition, American Mathematical Society, Colloquium Publications, Vol. XXIII, New York, USA.

[3] V. D. Pavlović, "New Class of Filter Functions Generated Directly by the Modified Christoffel–Darboux Formula for Classical Orthonormal Jacobi Polynomials", *International Journal of Circuit Theory and Applications*, Vol.41, pp.1059-1073, 2013.

[4] D. G. Ćirić, V. D. Pavlović, "Generalised Christoffel–Darboux Formula Most Directly Applied in Generating Fully Symmetric Doubly Resistively Terminated LC Lossless Ladder Filters", *International Journal of Electronics*, Vol.100, No.7, pp.942-958, 2013.

[5] A. Ilić, V. D. Pavlović, "New Class of Filter Functions Generated Most Directly by Christoffel-Darboux Formula for Gegenbauer Orthogonal Polynomials", *International Journal of Electronics*, Vol.98, No.1, pp.61–79, UK, 2011.

[6] V. D. Pavlović, A. Ilić, "New Class of Filter Functions Generated Most Directly by the Christoffel-Darboux Formula for Classical Orthonormal Jacobi Polynomials", *International Journal of Electronics*, Vol.98, No.12, pp.1603–1624, 2011.

[7] A. Ilic, V. D. Pavlović, "New Class of Filter Functions Generated Most Directly by Christoffel-Dorboux Formula for Gegenbauer Orthogonal

Polynomials", *International Journal of Electronics*, Vol.97, pp. 1-19, 2010.

[8] V. D. Pavlović, D. S. Antić, S. S. Nikolić, S. Lj. Perić, "Low complexity lowpass linear-phase multiplierless FIR filter", *Electronics Letters*, Vol.49, No.18, pp.1133-1135, 2013.

[9] S. Lj. Perić, D. S. Antić, V. D. Pavlović, S. S. Nikolić, M. T. Milojković, "Ultra Selective Low-pass Linear-phase FIR filter Function", *Electronics Letters,* Vol.49, No.9, pp.595-597, 2013.

[10] V. D. Pavlović, D. G. Ćirić, and N. S. Dončov, "Christoffel-Darboux Formula Most Directly Applied in Generating Economical Linear Phase Low-Pass Digital FIR Filter Functions", *Electronics and Electrical Engineering*, Vol.18, No.8, pp.109–112, 2012.

[11] V. D. Pavlović, D. S. Antić, S. Lj. Perić, S. S. Nikolić, M. T. Milojković, "Transitional Selective Linear Phase 1D FIR Filter Function Generated by Christoffel-Darboux Formula for Chebyshev Polynomials", *Electronics and Electrical Engineering*, in press, 2013.

[12] V. D. Pavlović, N. Dončov, and D. Ćirić, "1D and 2D FIR filters Generated by Chebyshev Polynomials of the First Kind", *International Journal of Electronics*, Vol.100, No.11, pp.1592-1619, 2013.

[13] J. R. Djordjević-Kozarov, V. D. Pavlović, "An Analytical Method for the Multiplierless 2D FIR Filter Functions and Hilbert Transform in z2 domain", *IEEE Transactions on Circuits and Systems- II: Express briefs,* Vol.60, No.8, pp.1-5, 2013.

[14] D. G. Ćirić, V. D. Pavlović, "Linear Phase Two-Dimensional FIR Digital Filter Functions Generated by applying Christoffel-Darboux Formula for Orthonormal Polynomials", *Electronics and Electrical Engineering*, Vol.120, No.4, pp.39-42, 2012.

[15] V. D. Pavlović, M. M. Lutovac, M. D. Lutovac, "Efficient Implementation of Multiplierless Recursive Lowpass FIR Filters using Computer Algebra System", *Proceedings of the 11th IEEE International Conference on Telecommunication in Modern Satellite Cable and Broadcasting Services (TELSIKS 2013)*, Vol.1, pp.65-68, 2013.

[16] M. D. Lutovac, V. D. Pavlovic, M. M. Lutovac "Efficient recursive implementation of multiplierless FIR filters", *Proceedings of the Mediterranean Conference on Embedded Computing (MECO 2013)*, pp.128-131, 2013.

[17] G. Jovanovic Dolecek, A. Fernandez-Vazquez, "Trigonometrical Approach to Design a Simples Wideband Comb Compensator",

International Journal of Electronics and Communications, accepted paper, paper no. AEUE-D-13-00497R1, 2014.

[18] J. O. Coleman, "Chebyshev stopbands for CIC decimation filters and CIC-implemented array tapers in 1D and 2D," *IEEE Trans. Circuits Syst-I: Reg. papers*, vol. 59, no. 12, pp. 2956–2968, Dec. 2012.

[19] G. Jovanovic Dolecek and A. Fernandez-Vazquez, "Novel droopcompensated comb decimation filter with improved alias rejections", *Int. Journal of Communications (AEÜ)*, vol. 67, pp. 387–396, 2013.

[20] G. Jovanovic Dolecek, "Simple wideband CIC compensator", *Electronics Letters*, vol. 45, no. 24, pp. 1270–1272, Nov. 2009.

[21] A. Kwentus, and A. Willson, Jr., "Application of filter sharpening to cascaded-integrator-comb decimation filters", *IEEE Trans. on Sig. Proc.*, vol. 45, no. 2, pp. 457-467, February 1997.

[22] S. C. Dutta Roy, "Impulse Response of sinc^N FIR Filters", *IEEE Transactions on Circuits and Systems*, vol. 53, no. 3, pp. 217–219, 2006.

INDEX

H

I

J

L

M

N

T

U

V

W

Y